PLATEAU'S PROBLEM
AND THE
CALCULUS OF VARIATIONS

PLATEAU'S PROBLEM
AND THE
CALCULUS OF VARIATIONS

by

Michael Struwe

Mathematical Notes 35

PRINCETON UNIVERSITY PRESS

PRINCETON, NEW JERSEY

1988

Printed in the United States of America by

Princeton University Press, 41 William Street,

Princeton, New Jersey 08540

The Princeton Mathematical Notes are edited by William Browder,

Robert Langlands, John Milnor, and Elias M. Stein

Library of Congress Cataloging in Publication Data

Struwe, Michael, 1955–
    Plateau's problem and the calculus of variations.
    (Mathematical notes ; 35)

    Bibliography: p.
    1. Surfaces, Minimal. 2. Plateau's problem.
3. Global analysis (Mathematics) 4. Calculus of
variations. I. Title. II. Series: Mathematical
notes (Princeton University Press) ; 35.
QA644.S77   1988   516.3'62   88-17963
ISBN 0-691-08510-2 (pbk.)

To Anne

Contents

Preface

Minimal surfaces and more generally surfaces of constant mean curvature - commonly known as soap films and soap bubbles - are among the oldest objects of mathematical analysis. The fascination that likewise attracts the mathematician and the child to these forms may lie in the apparent perfection and sheer beauty of these shapes. Or it may rest in the contrast between the utmost simplicity and endless variability of these remarkably stable and yet precariously fragile forms. The mathematician moreover may use soap films as a simple and beautiful model for his abstract ideas. In fact, long before the famous experiments of Plateau in the middle of the 19th century that initiated a first "golden age" in the mathematical study of minimal surfaces and through which Plateau's name became inseparably linked with these objects, Lagrange investigated surfaces of least area bounded by a given space curve as an illustration of the principle now known as "Euler-Lagrange equations". However, for a long time since Lagrange's derivation of the (non-parametric) minimal surface equation and Plateau's soap film experiments mathematicians had to acknowledge that their methods were completely inadequate for dealing with the Plateau problem in its generality. In spite of deep insights into the problem gained by applying the theory of analytic functions the solution to the classical Plateau problem evaded 19th century mathematicians- among them Riemann, Weierstraß, H.A. Schwarz. To meet the challenge ideas from complex analysis and the calculus of variations had to merge in the celebrated papers by J.Douglas and T. Radó in 1930/31.

But this was only the beginning of a new era of minimal surface theory in the course of which many significant contributions were made. Among other discoveries it was noted that the (parametric) Plateau problem may possess unstable solutions - which of course are not seen in the physical model - and in particular that the solutions to the Plateau problem in general are not unique. The question whether for "reasonable" boundary data the Plateau problem will always have only a finite number of solutions still puzzles mathematicians today. The most significant contributions are Tomi's result on the finiteness of the number of surfaces of absolutely minimal area spanning an analytic curve in $I\!R^3$ and the generic finiteness result of Böhme and Tromba.

The existence of "unphysical" solutions in the parametric problem in the 60's led to a new approach to the Plateau problem by what is now known as "geometric measure theory". In the course of these developments the notions of surface, area, tangent space, etc. came to be reconsidered and the notion of "varifold" evolved which parallels the notion of "weak solutions" in partial differential equations, cp. Nitsche [1, § 2].

But also the theory of the parametric Plateau problem was further pursued. Both to bring out the geometric content of the parametric solutions obtained (branch points, self-intersections) and - independently of the physical model - to explore the richness of a fascinating variational problem in its own right.

In this monograph we will focus our attention on the interplay between the parametric Plateau problem and developments in the calculus of variations, in particular global analysis. For reasons of space we will at most casually touch upon the more geometric aspects of the problem. As far as classical results about the geometry of minimal surfaces or the geometric measure theory approach to minimal surfaces are concerned the reader will find ample material and references in J.C.C. Nitsche's encyclopedic book [1] or in the lecture notes by L. Simon [1]. Our main emphasis will be on the power of the variational method.

Notations

    * denotes duality; occasionally we also denote a certain normalization with an asterisque.

These notes are divided into two parts with together four chapters, each divided into sections. Sections are numbered consecutively within each chapter. In cross-references to other chapters the number of the section is preceded by the number of the chapter which otherwise will be omitted.

These notes are based on lectures given at Louvain - La - Neuve and Bochum in 1985-86.

I am in particular indebted to Reinhold Böhme, Stefan Hildebrandt, Jean Mawhin, Anthony Tromba, Michel Willem and Eduard Zehnder for their continuous interest in the subject which has been a major stimulus for my work.

Special thanks I also owe to Herbert Gräff for his diligence and enthusiasm at typesetting this manuscript with the $\mathcal{A}_{\mathcal{M}}\mathcal{S}$-TEXsystem.

Finally, I wish to express my gratitude for the generous support of the SFB 72 at the University of Bonn.

Michael Struwe

Zürich, March 1988

PLATEAU'S PROBLEM
AND THE
CALCULUS OF VARIATIONS

A. The "classical" Plateau problem for

disc-type minimal surfaces

I. Existence of a solution.

1. The parametric problem. Let Γ be a Jordan curve in $I\!R^n$. The "classical" problem of Plateau asks for a disc-type surface X of least area spanning Γ. Necessarily, such a surface must have mean curvature 0. If we introduce isothermal coordinates on X (assuming that such a surface exists) we may parametrize X by a function $X(w) = (X^1(w), ..., X^n(w))$ over the disc

$$B = \left\{ w = (u,v) \in I\!R^2 \mid u^2 + v^2 < 1 \right\}$$

satisfying the following system of nonlinear differential equations

(1.1) $\triangle X = 0$ in B,

(1.2) $|X_u|^2 - |X_v|^2 = 0 = X_u \cdot X_v$ in B,

(1.3) $X|_{\partial B} : \partial B \to \Gamma$ is an (oriented) parametrization of Γ.

Here and in the following $X_u = \frac{\partial}{\partial u} X$, etc., and $\cdot$ denotes the scalar product in Euclidian $I\!R^n$.

Conversely, a solution to (1.1) - (1.3) will parametrize a surface of vanishing mean curvature (away from branch points where $\nabla X(w) = 0$) spanning the curve Γ, i.e. a surface satisfying the required boundary conditions and whose surface area is stationary in this class. Thus (1.1) - (1.3) may be considered as the Euler - Lagrange equations associated with Plateau's minimization problem.

However, (1.1) - (1.3) no longer require X to be absolutely area-minimizing. Correspondingly, in general solutions to (1.1) - (1.3) may have branch points, self-intersections, and be physically unstable - properties that we would not expect to observe in the soap film experiment. Thus as we specify the topological type of the solutions and relax our notion of "minimality" a new mathematical problem with its own characteristics evolves.

In the following we simply refer to solutions of (1.1) -(1.3) as minimal surfaces spanning Γ.

In this first chapter we present the classical solution to the parametric problem (1.1) - (1.3). Later we analyze the structure of the set of all solutions to (1.1) -(1.3). The key to this program is a variational principle for (1.1)-(1.3) which is "equivalent" to the least area principle but is not of a physical nature as it takes account of a feature present in the mathematical model but not in the physical solution itself: The parametrization of a solution surface. This variational principle is derived in the next section. Applying the "direct methods in the calculus of variations" we then obtain a (least area) solution to the problem of Plateau. At this stage the Courant-Lebesgue-Lemma will be needed. Finally, some results on the geometric nature of (least area) solutions will be recalled.

It will often be convenient to use complex notation and to identify points $w = (u, v) \in B$ with complex numbers $w = u + iv = re^{i\phi} \in \mathbb{C}$. Moreover, we introduce the complex conjugate $\overline{w} = u - iv$ and the complex differential operators

$$\partial = \frac{\partial}{\partial w} = \left(\frac{\partial}{\partial u} - i \frac{\partial}{\partial v} \right), \quad \overline{\partial} = \frac{\partial}{\partial \overline{w}} = \left(\frac{\partial}{\partial u} + i \frac{\partial}{\partial v} \right).$$

Note that $\overline{\partial}\partial = \triangle$; hence any solution X to (1.1) - (1.2) gives rise to a holomorphic differential $\partial X : B \subset \mathbb{C} \to \mathbb{C}^n$ satisfying the conformality relation $\partial X^2 = \sum_{i=1}^{n} \left(\partial X^i \right)^2 = 0$, cp. Lemma 2.3. Conversely, from any holomorphic curve $F : B \subset \mathbb{C} \to \mathbb{C}^n$ satisfying the compatibility conditon $F^2 = 0$ a solution $X(w) \equiv Real \int_{w_o}^{w} F \, dw$ to (1.1), (1.2) may be constructed.

This relation between minimal surfaces and holomorphic curves is the basis for the classical Weierstraß - Enneper representations of minimal surfaces in $I\!R^3$ which constitute one of the major tools for constructing and investigating minimal surfaces, cp. Nitsche [1, §§155 - 160] .

2. A variational principle.

Let $H^{1,2}(B; I\!R^n)$ be the Sobolev space of L^2-functions $X : B \to I\!R^n$ with square integrable distributional derivatives, and let

$$\|X\|_0^2 = \int_B |X|^2 dw$$

$$|X|_1^2 = \|\nabla X\|_o^2 = \int_B |\nabla X|^2 dw$$

$$\|X\|_1^2 = \|X\|_o^2 + |X|_1^2 = \int_B \left(|X|^2 + |\nabla X|^2\right) dw$$

denote the L^2-norm, respectively the seminorm and norm in $H^{1,2}(B; I\!R^n)$.

For $X \in H^{1,2}(B; I\!R^n)$ let

$$A(X) = \int_B \sqrt{|X_u|^2 |X_v|^2 - |X_u \cdot X_v|^2} dw$$

denote the area of the "surface" X, cp. Simon [1, p. 46].

Also introduce the class

$$\mathcal{C}(\Gamma) = \left\{ X \in H^{1,2}(B; I\!R^n) \mid \right.$$
$$\left. X|_{\partial B} \in C^0(\partial B, I\!R^n) \text{ is a weakly monotone parametrization of } \Gamma \right\}$$

of $H^{1,2}$-surfaces spanning Γ.

Note that the area of a surface X does not depend upon the parametric representation of X, i.e.

$$(2.1) \qquad A(X \circ g) = A(X)$$

for all diffeomorphisms g of $\overline{B}$. Hence by means of the area functional it is impossible to distinguish a particular parametrization of a surface X, and any attempt to approach the Plateau problem by minimizing A over the class $\mathcal{C}(\Gamma)$ is doomed to fail due to lack of compactness.

In 1930/31 Jesse Douglas and Tibor Radó however ingenuously proposed a different variational principle where the minimization-method meets success: They (essentially) considered Dirichlet's integral

$$D(X) = 1/2 \int_B |\nabla X|^2 dw$$

instead of A. For this functional the group of symmetries is considerably smaller; the relation

$$(2.2) \qquad D(X \circ g) = D(X)$$

only holds for conformal diffeomorphisms g of $\overline{B}$, i.e. for diffeomorphisms g satisfying the condition

(2.3) $$|g_u|^2 - |g_v|^2 = 0 = g_u \cdot g_v \quad \text{in } \overline{B}.$$

Now, A and D are related as follows: For $X \in H^{1,2}(B; I\!R^n)$

(2.4) $$A(X) \leq D(X)$$

with equality iff X is conformal, i.e. satisfies (1.2).

Conversely, given a surface parametrized by $X \in H^{1,2}(B; I\!R^n)$ we can assert the following result due to Morrey [2; Theorem 1.2]:

Theorem 2.1:   Let $X \in H^{1,2}(B; I\!R^n)$, $\epsilon > 0$. There exists a diffeomorphism $g : B \to B$ such that $X' = X \circ g$ satisfies:

$$D(X') \leq (1 + \epsilon)\, A(X') = (1 + \epsilon)\, A(X).$$

In particular, Theorem 2.1 implies that

(2.5) $$\inf_{X \in \mathcal{C}(\Gamma)} A(X) = \inf_{X \in \mathcal{C}(\Gamma)} D(X).$$

We will not prove Morrey's ϵ-conformality result. However, with the tools developed in Chapter 4 it will be easy to establish (2.5) for rectifiable Γ, cp. the appendix.

By (2.5), for the purpose of minimizing the area among surfaces in $\mathcal{C}(\Gamma)$ it is sufficient to minimize Dirichlet's integral in this class. Moreover, we have the following

Lemma 2.2: $X \in \mathcal{C}(\Gamma)$ solves the Plateau problem (1.1) - (1.3) iff X is critical for D on $\mathcal{C}(\Gamma)$ in the sense that

i) $\frac{d}{d\epsilon} D(X + \epsilon\varphi)|_{\epsilon=0} = 0, \ \forall \varphi \in H_o^{1,2}(B; I\!R^n)$

ii) $\frac{d}{d\epsilon} D(X \circ g_\epsilon^{-1}; B_\epsilon)|_{\epsilon=0} = 0$ for any family of diffeomorphisms $g_\epsilon : \overline{B} \to \overline{B}_\epsilon$ depending differentiably on a parameter $|\epsilon| < \epsilon_0$, and with $g_o = \text{id}$.

Proof :   Compute

$$\frac{d}{d\epsilon} D(X + \epsilon\varphi)|_{\epsilon=0} = \int_B \nabla X \nabla \varphi dw.$$

Hence the first stationarity condition i) is equivalent to the condition

$$\int_B \nabla X \nabla \varphi dw = 0, \quad \forall \varphi \in H_0^{1,2}(B; I\!R^n)$$

which in turn is just the weak form of the differential equation (1.1). By standard regularity results any weak solution $X \in H^{1,2}(B; I\!\!R^n)$ of (1.1) will be smooth in B and (1.1) will be satisfied in the classical sense.

It remains to show that for harmonic $X \in \mathcal{C}(\Gamma)$ the stationarity condition ii) is equivalent to the conformality relations (1.2). This result requires some preparatory lemmata which we state in a slightly more general way than will actually be needed.

Lemma 2.3: Let G be a domain in $I\!\!R^2 \stackrel{\sim}{=} \mathbb{C}$ and suppose $X \in H^{1,2}(G; I\!\!R^n)$ is harmonic. Then the function

$$\Phi(w) \equiv |X_u|^2 - |X_v|^2 - 2i\, X_u \cdot X_v$$

is a holomorphic function of $w = u + iv \in G \subset \mathbb{C}$.

Proof: Note that Φ may be written as a product

$$\Phi = (X_u - iX_v)^2 = (\partial X)^2$$

with component-wise complex multiplication and

$$\partial = \left(\frac{\partial}{\partial u} - i \frac{\partial}{\partial v} \right), \quad \overline{\partial} = \left(\frac{\partial}{\partial u} + i \frac{\partial}{\partial v} \right)$$

the usual complex differential operators.

Note that

$$\overline{\partial}\partial = \frac{\partial^2}{\partial u^2} + \frac{\partial^2}{\partial v^2} = \triangle.$$

Hence by harmonicity of X

$$\overline{\partial}\Phi = 2\overline{\partial}\partial X \cdot \partial X = 2\triangle X \cdot \partial X = 0,$$

i.e. Φ is holomorphic.

□

Lemma 2.4: Suppose G is a domain in $I\!\!R^2$, and let $X \in H^{1,2}(G; I\!\!R^n)$. Moreover, suppose that for any differentiable family of diffeomorhisms $g_\epsilon : \overline{G} \to \overline{G}_\epsilon$ with $g_o = \text{id}$ there holds

$$\frac{d}{d\epsilon} D\left(X \circ g_\epsilon^{-1};\ G_\epsilon\right)|_{\epsilon=0} = 0.$$

Then X is conformal.

Proof: Let $\tau \in C^1(\overline{G}; I\!\!R^2)$ and for $\epsilon \in I\!\!R$ with

$$|\epsilon|\, \|\nabla\tau\|_{L^\infty} < 1$$

consider maps $g_\epsilon = id + \epsilon\tau : G \to G_\epsilon := g_\epsilon(G)$. Since by choice of ϵ the maps g_ϵ are injective and the rank of the differential

$$\nabla g_\epsilon = id + \epsilon\nabla\tau$$

is maximal everywhere the g_ϵ in fact are diffeomorphisms $g_\epsilon : \overline{G} \to \overline{G}_\epsilon$.

Compute by the chain rule:

$$
\begin{aligned}
D(X \circ g_\epsilon^{-1}; G_\epsilon)|_{\epsilon=0} &= \frac{1}{2}\int\limits_{G_\epsilon} \left|\nabla(X \circ g_\epsilon^{-1})\right|^2 dw \\
&= \frac{1}{2}\int\limits_{G_\epsilon} \left|((\nabla X) \circ g_\epsilon^{-1}) \cdot \nabla g_\epsilon^{-1}\right|^2 dw \\
&= \frac{1}{2}\int\limits_{G} \left|\nabla X \cdot ((\nabla g_\epsilon^{-1}) \circ g_\epsilon)\right|^2 \ \det(\nabla g_\epsilon)dw
\end{aligned}
$$

Now $g_\epsilon^{-1} \circ g_\epsilon = $ id implies that

$$(\nabla g_\epsilon^{-1}) \circ g_\epsilon = (\nabla g_\epsilon)^{-1} = \text{ id } - \epsilon\nabla\tau + 0(\epsilon)^2$$

while - labeling $\tau = (\tau^1, \tau^2)-$

$$\det(\nabla g_\epsilon) = 1 + \epsilon(\tau_u^1 + \tau_v^2) + 0(\epsilon^2).$$

I.e.

$$
\begin{aligned}
D(X \circ g_\epsilon^{-1}; G_\epsilon) = 1/2 \int\limits_{G} &|\nabla X|^2 - 2\epsilon\left(|X_u|^2\tau_u^1 + |X_v|^2\tau_v^2 + X_u \cdot X_v\,(\tau_u^2 + \tau_v^1)\right) \\
&+ \epsilon|\nabla X|^2(\tau_u^1 + \tau_v^2)dw + 0(\epsilon^2).
\end{aligned}
$$

It is now clear that $\epsilon \mapsto D(X \circ g_\epsilon^{-1}; G_\epsilon)$ is differentiable at $\epsilon = 0$ and

$$
\begin{aligned}
\frac{d}{d\epsilon}D(X \circ g_\epsilon^{-1}&; G_\epsilon)|_{\epsilon=0} = \\
&= -\frac{1}{2}\int\limits_{G}(|X_u|^2 - |X_v|^2)(\tau_u^1 - \tau_v^2) + 2\ X_u \cdot X_v(\tau_u^2 + \tau_v^1)du\,dv\ .
\end{aligned}
$$

If now again we consider $I\!R^2 \hat{=} \mathbb{C}$ by letting $w = u + iv$, $\tau = \tau^1 + i\tau^2$ we may rewrite the integrand as follows:

$$(|X_u|^2 - |X_v|^2)(\tau_u^1 - \tau_v^2) + 2\ X_u \cdot X_v(\tau_v^1 + \tau_u^2) = Real\ (\Phi \cdot \overline{\partial}\tau),$$

where Φ is defined as in Lemma 2.3.

Thus

(2.6) $$\frac{d}{d\epsilon}D(X \circ g_\epsilon^{-1}; G_\epsilon)|_{\epsilon=0} = -1/2\int\limits_{G} Real(\Phi \cdot \overline{\partial}\tau)du\,dv$$

and the expression can only vanish for all $\tau \in C^1(\overline{G}; I\!\!R^2)$ if Φ vanishes identically in G, i.e. if X is conformal.

□

To conclude the **proof of Lemma 2.2** in view of Lemma 2.4 it suffices to remark that by (2.6) conformality of X also implies the stationarity condition ii) of Lemma 2.2. Hence the critical points of D in $\mathcal{C}(\Gamma)$ precisely correspond to the solutions of Plateau's problem.

□

Remarks 2.5. i) If X is harmonic on B, by Lemma 2.3 and upon integrating by parts in (2.6) we obtain

$$\frac{d}{d\epsilon}D(X \circ g_\epsilon^{-1}; B_\epsilon)\,|_{\epsilon=0} = -\frac{1}{2}\int\limits_{\partial B} Real\,(\Phi \cdot w\tau)\ do.$$

Thus, the conformality relations (1.2) may be interpreted as a natural boundary condition for the holomorphic function Φ associated with X. Cp. Courant [1, p. 72 ff].

ii) Variations of the type i) in Lemma 2.1 may be interpreted as "variations of the dependent variables" i.e. of the *surface* X. Variations of the type ii) ("variations of the independent variables") correspond to variations of the *parametrization* of X.

iii) By conformal invariance of D and the Riemann mapping theorem any minimizer X_0 of D in $\mathcal{C}(\Gamma)$ will be a critical point of D in the sense of Lemma 2.1. Indeed, by (2.6) it suffices to show that X_0 satisfies the stationarity condition ii) of Lemma 2.2 for all $g_\epsilon = \text{id} + \epsilon\tau, \tau \in C^1(\overline{B}; I\!\!R^2)$.

Suppose by contradiction that for some $\tau \in C^1(\overline{B}; I\!\!R^2)$

$$\frac{d}{d\epsilon}D(X_0 \circ (\text{id} + \epsilon\tau)^{-1}, B_\epsilon)|_{\epsilon=0} \neq 0,$$

with $B_\epsilon = (\text{id} + \epsilon\tau)(B)$. Then for some $\epsilon \neq 0$ and $X_\epsilon = X_0 \circ (\text{id} + \epsilon\tau)^{-1}$ we have

$$D(X_\epsilon; B_\epsilon) < D(X_0).$$

But B_ϵ is conformal to B. Hence we may compose X_ϵ with a conformal map $g_\epsilon : \overline{B} \to \overline{B_\epsilon}$ to obtain a comparison surface $\tilde{X}_\epsilon = X_\epsilon \circ g_\epsilon \in \mathcal{C}(\Gamma)$ with

$$D(\tilde{X}_\epsilon) = D(X_\epsilon; B_\epsilon) < D(X_0) = \min\{D(X)|X \in \mathcal{C}(\Gamma)\}.$$

The contradiction proves that X_0 is critical for D.

3. The direct methods in the calculus of variations.   We now proceed to derive the existence of a minimizer of D on $\mathcal{C}(\Gamma)$-and hence of a solution to Plateau's problem (1.1) - (1.3), cp. Remark 2.4. iii) - from the following general principle:

Theorem 3.1:   Let M be a topological Hausdorff space, and let $E : M \to \mathbb{R} \cup \{\infty\}$.

Suppose that for any $\alpha \in \mathbb{R}$ the set

(3.1) $$\overline{M_\alpha} := \{ x \in M \mid E(x) \leq \alpha \}$$

is compact.

Then there exists $x_0 \in M$ such that

$$E(x_0) = \inf_{x \in M} E(x).$$

In particular, E is bounded from below and lower semi-continuous on M.

Proof: Let

$$\alpha_0 = \inf_{x \in M} E(x) \geq -\infty,$$

and consider a sequence $\{\alpha_m\}$ of numbers $\alpha_m > \alpha_0$ tending to α_0 as $m \to \infty$.

By compactness of $\overline{M_{\alpha_1}}$ for $\alpha \in \mathbb{R}$, the nested sequence

$$\overline{M_{\alpha_1}} \supset \overline{M_{\alpha_2}} \supset \ ...$$

has non-empty intersection and there exists

$$x_0 \in \bigcap_{m \in \mathbb{N}} \overline{M_{\alpha_m}}.$$

Clearly, $\alpha_0 \leq E(x_0) \leq \alpha_m$ for any m and therefore letting $m \to \infty$ we infer that

$$E(x_0) = \alpha_0 .$$

Since E does not assume the value $-\infty$, in particular $\alpha_0 > -\infty$, and E is bounded from below on M.

Finally, by (3.1) for any $\alpha \in \mathbb{R}$ the set $\{x \in M \mid E(x) > \alpha\}$ is open, i.e. E is lower semi-continuous.

□

Remark 3.2:   In the work of M. Morse compactness of the sets $\overline{M}_\alpha$ in Theorem 3.1 defines the property of "bounded compactness" of E on M. This condition implies lower semi-continuity of E. However E cannot be continuous on M and simultaneously satisfy (3.1) unless M is locally compact: By (3.1) any set

$$M_\alpha = \{x \in M \mid E(x) < \alpha\}$$

must be relatively compact in M while by continuity M_α is also open.

In applications a simple variant of Theorem 3.1 will often be encountered:

Theorem 3.3:   Suppose M is a sub-set of a separable Hilbert space H which is closed with respect to the weak topology on H.

Let $E : M \to I\!R$ be a funtional which is *sequentially weakly lower semi-continuous* on M, i.e. which satisfies the condition

$$(3.2) \qquad E(x) \leq \lim_{m \to \infty} \inf E(x_m), \quad \text{whenever } x_m \xrightarrow{w} x, x_m \in M.$$

Also assume that E is *coercive*, i.e. suppose that for any sequence $\{x_m\}$ in M there holds:

$$(3.3) \qquad E(x_m) \to \infty \text{ whenever } |x_m|_H \to \infty.$$

Then there exists a minimizer $x_0 \in M$ with

$$E(x_0) = \inf_{x \in M} E(x).$$

Theorem 3.3 is reduced to Theorem 3.1 by letting M be endowed with the weak topology on H. However, there also is a very natural direct and constructive proof of Theorem 3.3 which uses the concept of a *minimizing sequence*.

Proof of Theorem 3.3:   Let

$$\alpha_0 = \inf_{x \in M} E(x) \geq -\infty,$$

and let $\{x_m\} \subset M$ be a sequence such that $E(x_m) \to \alpha_0$ as $m \to \infty$. By coerciveness of E $\{x_m\}$ is bounded and hence weakly relatively compact. Extracting a weakly convergent subsequence $x_m \xrightarrow{w} x_0$, by weak closedness of M also the weak limit $x_0 \in M$. Finally, by (3.2)

$$\alpha_0 \leq E(x_0) \leq \lim_{m \to \infty} \inf E(x_m) = \alpha_0,$$

and the proof is complete.

$\square$

Examples 3.4:   i) the norm in a Hilbert space H with scalar product $(\cdot, \cdot)$

$$|x|^2 = (x, x)$$

is weakly lower semi-continuous.

ii)   More generally, let $a : H \times H \to I\!R$ be a continuous symmetric bilinear form on H such that

$$a(x, x) \geq 0, \quad \forall \, x \in H.$$

Then $E(x) \equiv a(x, x)$ is weakly lower semi-continuous on H.

In particular, D is weakly lower semi-continuous on $H^{1,2}(B; I\!R^n)$.

Proof: Suppose $x_m \overset{w}{\rightharpoonup} x$. Then

$$0 \leq a(x_m - x, \; x_m - x) \; = \; a(x_m, x_m) - a(x, x) - 2a(x, x_m - x).$$

By the Riesz representation theorem there exists $y \in H$ such that

$$a(x, x_m - x) \; = \; (y, \; x_m - x) \to 0.$$

$\square$

iii) Suppose $E : H \to I\!R \cup \{\infty\}$ is continuous and *convex*, i.e. for all $x, y \in H$ $0 < \alpha < 1$ there holds

(3.4) $$E(\alpha x + (1 - \alpha)y) \; \leq \; \alpha E(x) \; + \; (1 - \alpha)E(y).$$

Then E is weakly lower semi-continuous on H.

Proof: If $x_m \overset{w}{\rightharpoonup} x$ weakly, by the Banach-Saks theorem

$$x^N := \frac{1}{N} \sum_{m=1}^{N} x_m \to x \quad \text{strongly in} \; H \; \text{as} \; N \to \infty.$$

Hence, by continuity of E and (3.4)

$$E(x) \; = \; \lim_{N \to \infty} E(x^N) \leq \lim_{N \to \infty} \frac{1}{N} \sum_{m=1}^{N} E(x_m) \; \leq \; \liminf_{m \to \infty} \; E(x_m).$$

Remark: The inequality

$$E\left(\frac{1}{N}\sum_{m=1}^{N} x_m\right) \leq \frac{1}{N}\sum_{m=1}^{N} E(x_m), \quad \{x_m\} \subset H, \ N \in I\!N$$

for a convex functional $E : H \to I\!R \cup \{\infty\}$ is a special case of *Jensen's* inequality.

Example 3.5: The functional D is coercive on $\mathcal{C}(\Gamma)$.

Proof: By the Sobolev inequality for $X \in H^{1,2}(B; I\!R^n)$ with $X|_{\partial B} \in L^\infty(\partial B; I\!R^n)$:

$$\|X\|_o^2 \leq c \left(\|\nabla X\|_o^2 + \|X\|_{o,\partial B}^2\right) \leq c \left(D(X) + \|X\|_{L^\infty(\partial B)}^2\right).$$

Hence for $X \in \mathcal{C}(\Gamma)$: $\|X\|_1^2 \leq c\, D(X) + c(\Gamma)$.

4. The Courant Lebesgue Lemma and its consequences.

In the preceding chapter we have seen the importance of weak closedness of $\mathcal{C}(\Gamma)$. However, the presence of the conformal group of the disc

$$(4.1) \qquad G = \left\{ g(w) = e^{i\phi_0} \frac{a+w}{1+\overline{a}w} \;\middle|\; a \in \mathbb{C},\ |a| < 1,\ \phi_0 \in \mathbb{R} \right\}$$

acting on $\mathcal{C}(\Gamma)$ and conformal invariance of D cause problems.

Lemma 4.1:   For $X \in \mathcal{C}(\Gamma)$ let $X \circ G = \{X \circ g \mid g \in G\}$ be the conformal orbit of X. Then for any X the weak closure of $X \circ G$ contains a constant map.

Proof: i) First consider $\varphi \in C^1(\overline{B}; \mathbb{R}^n)$. Let $g_m(w) = \dfrac{a_m + w}{1 + \overline{a}_m w}$, where $a_m \in \mathbb{C}$, $|a_m| < 1$, $a_m \to 1$.

Clearly, as $m \to \infty$

$$g_m(w) \to 1$$

for all $w \in B$, uniformly away from $w = -1$.

Hence

$$\varphi_m = \varphi \circ g_m \to \varphi(1)$$

pointwise in B.

By conformal invariance of D moreover

$$D(\varphi_m) = D(\varphi) < \infty$$

while

$$|\varphi_m|_{L^\infty} = |\varphi|_{L^\infty} < \infty,$$

and $\{\varphi_m\}$ admits also a weakly convergent subsequence $\varphi_m \overset{w}{\to} \varphi_0 \equiv \varphi(1)$. This proves our claim for regular functions.

ii) For $X \in \mathcal{C}(\Gamma)$ let g_m be as above and define

$$X_m = X \circ g_m^{-1} \in \mathcal{C}(\Gamma).$$

By (2.2) and Example 3.5 $\{X_m\}$ is bounded in $H^{1,2}(B; \mathbb{R}^n)$ and we may extract a subsequence $X_m \overset{w}{\to} X_0$.

To show that $X_0 \equiv \text{const}$ it suffices to show that

$$\int_B \nabla X_0 \nabla \varphi \cdot dw = 0, \ \forall \ \varphi \in C^1(\overline{B}; \mathbb{R}^n).$$

But by i) of this proof and conformal invariance of D, with $\varphi_m = \varphi \circ g_m$ we have :

$$\int_B \nabla X_0 \nabla \varphi \, dw = \lim_{m \to \infty} \int_B \nabla X_m \nabla \varphi \, dw$$

$$= \lim_{m \to \infty} \int_B \nabla X \nabla \varphi_m \, dw = 0,$$

which completes the proof.

□

In view of Lemma 4.1 the set $\mathcal{C}(\Gamma)$ cannot be weakly closed in $H^{1,2}(B; I\!R^n)$. However, equivariance of D with respect to G allows us to factor out the symmetry group. The most convenient way to do this is by imposing a three - point - condition on admissible functions. Note that (4.1) immediately implies:

Lemma 4.2:   Given any triples (ϕ_1, ϕ_2, ϕ_3), (ψ_1, ψ_2, ψ_3), $0 \leq \varphi_1 < \varphi_2 < \varphi_3 < 2\pi$, $0 \leq \psi_1 < \psi_2 < \psi_3 < 2\pi$,   there exists a unique $g \in G$ such that

$$g(e^{i\varphi_j}) = e^{i\psi_j}, \ j = 1, 2, 3.$$

Lemma 4.2 suggests to normalize admissible functions as follows: Let $P_j = e^{\frac{2\pi i j}{3}}, j = 1, 2, 3$ and let $Q_j, \ j = 1, 2, 3$ be an oriented triple of distinct points on Γ. Define

$$C^*(\Gamma) = \{X \in \mathcal{C}(\Gamma) | X(P_j) = Q_j, \ j = 1, 2, 3\}.$$

Then we obtain the following crucial result:

Lemma 4.3   The injection $C^*(\Gamma) \rightarrow C^0(\partial B; I\!R^n)$ is compact, i.e. $D-$ bounded subsets of $C^*(\Gamma)$ are equicontinuous on ∂B.

For the proof we need the following fundamental lemma due to Courant [1, p. 101 ff.] and Lebesgue [1, p. 388]:

Lemma 4.4:   For any $X \in H^{1,2}(B; I\!R^n)$, any $w \in \overline{B}$, any $\delta \in]0, 1[$ there exists $\rho \in [\delta, \sqrt{\delta}]$ such that if s denotes   arc length on

$$C_\rho = C_\rho(w) = \partial B_\rho(w) \cap B$$

we have: $X_s \in L^2(C_\rho)$ and

$$\int\limits_{C_\rho} |X_s^2| ds \ \leq \ 8D(X)/\rho|\ln \rho|.$$

Proof: By Fubini's theorem $|X_s| \in L^2(C_\rho)$ for a.e. $\rho < 1$ and

$$2D(X) \geq \int\limits_{(B_{\sqrt{\delta}}(w) \setminus B_\delta(w)) \cap B} |\nabla X|^2 dw \geq \int\limits_{\delta}^{\sqrt{\delta}} \int\limits_{C_\rho} |X_s|^2 ds \, d\rho$$

$$\geq \operatorname*{essinf}_{\delta \leq \rho \leq \sqrt{\delta}} \left(\rho \int\limits_{C_\rho} |X_s|^2 ds \right) \cdot \int\limits_{\delta}^{\sqrt{\delta}} \frac{d\rho}{\rho}.$$

Since for all $\rho \in [\delta, \sqrt{\delta}]$

$$\int\limits_{\delta}^{\sqrt{\delta}} \frac{d\rho}{\rho} \ = \ 1/2 \, |\ln \delta| \ \geq 1/2 \, |\ln \rho|,$$

we can find ρ as claimed.

□

Proof of Lemma 4.3:   Let $X \in C^*(\Gamma)$, $\epsilon > 0, w_0 \in \partial B$. We contend that there exists a number $\delta > 0$ depending only on ϵ , $D(X)$, the curve Γ, and the points $Q_j, 1 \le j \le 3$, such that for all $w \in \partial B$ there holds

$$(4.2) \qquad |X(w) - X(w_0)| < 2\epsilon \text{ , if } |w - w_0| < \delta.$$

This statement is equivalent to the contended equicontinuity of D-b(•nded subsets of $C^*(\Gamma)$. By a theorem of Arzéla - Ascoli the latter in turn is equivalent to the compactness of the injection $C^*(\Gamma) \to C^0(\partial B; I\!\!R^n)$.

Choose $\delta_0 > 0$ small enough such that any ball of radius $\sqrt{\delta_0}$ contains at most one of the points $P_j = e^{\frac{2\pi i j}{3}}, j = 1, 2, 3$. Choose $\epsilon_o > 0$ such that a ball of radius ϵ_0 in $I\!\!R^n$ contains at most one of the three points Q_j, $j = 1, 2, 3$. Clearly, we may assume that $\epsilon < \epsilon_0$. Choose ϵ_1, $0 < \epsilon_1 < \epsilon$, such that for any two points X, Y on Γ at a distance $|X - Y| < \epsilon_1$ there is a subarc $\tilde{\Gamma} \subset \Gamma$ with end-points X and Y contained in some ball of radius ϵ in $I\!\!R^n$. (This is possible for any Jordan curve Γ. Otherwise, for sequences $\{X_m\}, \{Y_m\}$ of points in Γ with $|X_m - Y_m| \to 0$ $(m \to 0)$ any subarc $\tilde{\Gamma}_m$ joining X_m with Y_m would intersect $\partial B_\epsilon(X_m)$ in a point Z_m. By compactness of Γ we may assume $X_m \to X$, $Y_m \to Y = X$, $Z_m \to Z$, $|X - Z| = \epsilon$. In particular, the limits Y and X correspond to different parameter values of a given parametrization of Γ. But this contradicts our assumption that Γ is a Jordan curve, i.e. a homeomorphic image of S^1.)

By choice of ϵ_o, for $X, Y \in \Gamma$ with $|X - Y| < \epsilon_1$ the subarc $\tilde{\Gamma} \subset \Gamma$ connecting X and Y and lying in a ball of radius ϵ in $I\!\!R^n$ is unique and is characterized by the condition that $\tilde{\Gamma}$ contains at most one the points Q_j, $j = 1, 2, 3$.

Now choose a maximal δ, $0 < \delta \le \delta_0$, such that

$$|\ln \delta| \ge \frac{16\pi D(X)}{\epsilon_1^2} \ .$$

Let $\rho \in [\delta, \sqrt{\delta}]$, $C_\rho = C_\rho(w_0)$ be selected according to Lemma 4.4 satisfying

$$\int_{C_\rho} |X_s|^2 ds \le \frac{8D(X)}{\rho |\ln \rho|} \ .$$

Denote $w_j = e^{2\pi i \phi_j}$, $j = 1, 2$, the points of intersection of C_ρ with ∂B, $\tilde{C}_\rho = \partial B \cap B_\rho(w_o)$ that subarc of ∂B with end-points w_1, w_2 which contains at most one of the points P_j, $j = 1, 2, 3$. Also let $X_j = X(w_j)$, $j = 1, 2$, and let $\tilde{\Gamma}$ be that subarc of Γ connecting X_1, X_2 containing at most one of the points $Q_j, j = 1, 2, 3$. By monotonicity $X(\tilde{C}_\rho) = \tilde{\Gamma}$. Moreover, by Hölder's inequality:

$$|X_1 - X_2|^2 \le \left(\int_{C_\rho} |X_s| ds \right)^2 \le \pi \rho \int_{C_\rho} |X_s|^2 ds$$

$$\le 8\pi \, D(X)/|\ln \rho| \le 16\pi \, D(X)/|\ln \delta| \le \epsilon_1^2.$$

By choice of ϵ_1, $\tilde{\Gamma}$ is contained in a ball of radius ϵ. In particular, for w_o and any $w \in \partial B \cap B_\delta(w_o) \subset \tilde{C}_\rho$ there holds

$$|X(w) - X(w_o)| \leq 2\epsilon$$

Since δ depends only on $D(X)$ and ϵ_1 while the latter only depends on ϵ, Γ, Q_j, $j = 1, 2, 3$, the proof is complete.

□

Lemma 4.3 immediately implies the following results:

Proposition 4.5: The set $C^*(\Gamma)$ is closed with respect to the weak topology in $H^{1,2}(B; I\!R^n)$.

Proof: Consider a sequence $\{X_m\} \subset C^*(\Gamma)$ such that $X_m \overset{w}{\to} X$ weakly in $H^{1,2}(B; I\!R^n)$. By weak convergence, $\{X_m\}$ is bounded and in particular for some $c \in I\!R$

$$D(X_m) \leq c$$

uniformly in m. Lemma 4.3 now implies that (a subsequence) $X_m \to X$ uniformly on ∂B. Hence $X \in C^*(\Gamma)$, and $C^*(\Gamma)$ is weakly $H^{1,2}$-closed.

□

Together with coercivenes of D on $C(\Gamma)$ (cp. Example 3.5) and weak lower semi-continuity of D on $H^{1,2}(B; I\!R^n)$ (cp. Example 3.4. ii)) Proposition 4.5 implies:

Proposition 4.6: Suppose Γ is a Jordan curve in $I\!R^n$ such that $C(\Gamma) \neq \emptyset$. Then there exists a solution to Plateau's problem (1.1) - (1.3) parametrizing a minimal surface of disc-type spanning Γ.

Proof: Indeed, let $C^*(\Gamma)$ be defined as above with reference to a conveniently chosen triple (Q_1, Q_2, Q_3) of points on Γ. Theorem 3.3 guarantees the existence of a surface $\underline{X} \in C^*(\Gamma)$ such that

$$D(\underline{X}) = \inf_{X \in C^*(\Gamma)} D(X).$$

Moreover, for any $X \in C(\Gamma)$ by Lemma 4.2 there is a unique conformal diffeomorphism g of B such that $X' = X \circ g \in C^*(\Gamma)$. By conformal invariance of D also $D(X') = D(X)$, and it follows that

$$\inf_{X \in C^*(\Gamma)} D(X) = \inf_{X \in C(\Gamma)} D(X).$$

Consequently, $\underline{X}$ minimizes D over $C(\Gamma)$. Hence by Remark 2.5.iii) $\underline{X}$ furnishes a solution to the parametric form (1.1) - (1.3) of Plateau's problem.

□

For later reference we also note the following compactness result:

Proposition 4.7:   Suppose Γ is a Jordan curve in $I\!R^n$. Then for any $\beta \in I\!R$ the set

$$\{X \in \mathcal{C}^*(\Gamma) \mid D(X) \leq \beta\}$$

is compact with respect to the weak topology in $H^{1,2}(B; I\!R^n)$ and the $C^0(\partial B; I\!R^n)$ -topology of uniform convergence on ∂B.

Proposition 4.7 is easily deduced from Proposition 4.5 using the coerciveness and weak lower semi-continuity of D on $\mathcal{C}^*(\Gamma)$. Combining Proposition 4.7 and Theorem 3.1 would give an alternative proof of Proposition 4.6.

It remains to give a general condition for $\mathcal{C}(\Gamma)$ to be non-void. This is contained in

Lemma 4.8:   For any rectifiable Jordan curve $\Gamma \subset I\!R^n$ the class $\mathcal{C}(\Gamma) \neq \emptyset$.

Our proof rests on the following a-priori bound for the area of solutions to (1.1) - (1.3):

Theorem 4.9   (Isoperimetric inequality):   Suppose Γ is a rectifiable Jordan curve in $I\!R^n$ with length $L(\Gamma) < \infty$. Then for any solution $X \in \mathcal{C}(\Gamma)$ of (1.1) - (1.3) there holds the estimate

$$4\pi\, D(X) \leq (L(\Gamma))^2.$$

The constant 4π is best possible.

Cp. Nitsche [1, §323].

For our purposes it will be sufficient to establish the qualitative bound

(4.3) $$D(X) \leq c(L(\Gamma))^2$$

for any $C^1(\overline{B}; I\!R^n)$-solution to (1.1) - (1.3).

Proof of (4.3):   Multiply (1.1) by X and integrate by parts to obtain

$$2D(X) = \int_B |\nabla X|^2 dw = \int_{\partial B} \partial_n X \cdot X \, do$$

$$\leq \int_{\partial B} |\partial_n X||X| do \leq \|\Gamma\|_{L^\infty} \int_{\partial B} |\partial_\tau X| do,$$

where do denotes the one-dimensional measure on ∂B, and n and τ are unit normal and tangent vector fields to ∂B. Of course, by (1.3)

$$\int_{\partial B} |\partial_\tau X| do = L(\Gamma),$$

while by suitable choice of coordinates in $I\!R^n$ such that $0 \in \Gamma$

$$\|\Gamma\|_{L^\infty} \leq L(\Gamma)/2.$$

This proves (4.3) with $c = \frac{1}{4}$.

□

Proof of Lemma 4.8:   Approximate Γ by smooth Jordan curves Γ_m in $I\!R^{n+2}$ of class C^2 on $\partial B \hat{=} I\!R/2\pi$. This can be done as follows: Let $\gamma \in H^{1,1}(\partial B; I\!R)$ be a homeomorphism $\gamma : \partial B \rightarrow \Gamma$. First convolute γ with a sequence $\{\tau_m\}$ of non-negative $\tau_m \in C^\infty(I\!R)$ vanishing for $|\phi| \geq \frac{1}{m}$ and satisfying $\int \tau_m(\phi) d\phi = 1$ to obtain a sequence of smooth maps $\tilde{\gamma}_m(\phi) \equiv \int_{I\!R} \gamma(\phi - \phi') \tau_m(\phi') d\phi'$. Then let $\gamma_m(\phi) = (\tilde{\gamma}_m(\phi), \frac{1}{m} e^{i\phi})$, $\Gamma_m = \gamma_m(\partial B)$. In this way we generate a sequence of C^2-diffeomorphisms such that

$$\gamma_m \rightarrow \gamma \quad \text{in} \quad H^{1,1}(\partial B; I\!R^{n+2}).$$

Extending the parametrizations γ_m to harmonic surfaces X_m we immediately see that $\mathcal{C}(\Gamma_m) \neq \emptyset$ for all m. By Proposition 4.6 there exist solutions $X_m \in \mathcal{C}(\Gamma_m)$ to (1.1) - (1.3) for Γ_m, moreover, by Theorem 5.1 below $X_m \in C^1(\overline{B}; I\!R^{n+2})$. But then Theorem 4.9 assures the bound

$$(4.4) \qquad\qquad D(X_m) \leq c \left(L(\Gamma_m)\right)^2 \leq c(\Gamma)$$

for large m.

Now let $Q_j = \gamma(P_j)$, $Q_j^{(m)} = \gamma_m(P_j)$, $j = 1, 2, 3$, $m \in I\!N$. With no loss of generality we may assume that $X_m \in \mathcal{C}^*(\Gamma_m)$ where we normalize with reference to the triples $(Q_j^{(m)}), j = 1, 2, 3$. Since for this normalization the constants $\epsilon_1^{(m)}$ appearing in the proof of Lemma 4.3 clearly have a uniform lower bound $\epsilon_1 > 0$, (4.4) and the proof of Lemma 4.3 show that the surfaces X_m are equicontinuous on ∂B. Hence a subsequence $X_m \xrightarrow{w} X$ weakly in $H^{1,2}(B; I\!R^{n+2})$ and uniformly on ∂B. Note that X is harmonic with $X(\partial B) = \Gamma \subset I\!R^n$. Thus by the maximum principle $X \in H^{1,2}(B; I\!R^n)$ and $X \in \mathcal{C}(\Gamma)$.

□

Lemma 4.8 and Proposition 4.6 finally yield the following existence result of Douglas [1] and Radó [1]:

Theorem 4.10:   Let Γ be a rectifiable Jordan curve in $I\!R^n$. Then there exists a solution $\underline{X}$ to (1.1) - (1.3) characterized by the condition

$$D(\underline{X}) = \inf_{X \in \mathcal{C}(\Gamma)} D(X) < \infty,$$

and $\underline{X}$ parametrizes a disc-type minimal surface of least area spanning Γ.

5. Regularity.   The preceding considerations establish the existence of a solution
to the parametric Plateau problem (1.1) - (1.3). In order to interpret this solution
geometrically we now derive further regularity properties. Note that the regularity
question is two-fold: First we analyze the regularity of the parametrization; then
we turn to the question whether the parametrized surface is regular enough to be
admitted as a solution to Plateau's problem, i.e. whether it is embedded (or at least
locally immersed).

While the first question is completely solved by Hildebrandt's regularity result [1] ,
for the second question a satisfactory answer can only be given in case $n = 3$ which
corresponds to the physical case. In this case the results of Osserman [1] , Alt [1],
Gulliver [1], Gulliver, Osserman, and Royden [1], Gulliver and Lesley [1], Sasaki
[1] and Nitsche [1, p. 346] show that the solutions of Douglas and Radó will be
free of interior branch points and hence will be immersed over B−and even be
immersed over $\overline{B}$ if Γ is analytic or has total curvature $\leq 4\pi$. For extreme
curves, i.e.   curves on the boundary of a region $\Omega \subset \mathbb{R}^3$ which is convex or
more generally whose boundary has non-negative mean curvature with respect to
the interior normal, Meeks and Yau [1] even have proved that a least area solution to
(1.1) - (1.3) parametrizes an embedded minimal disc. Related results were obtained
independently by Tomi and Tromba [1] , resp. Almgren and Simon [1] . This
extends an old result of Radó [2] for curves having a single valued parallel projection
onto a convex planar curve. Simple examples show that without such additional
geometric conditions on Γ in general least-area solutions to (1.1) - (1.3) need not
be embedded.

Below we briefly survey some of the most significant contributions to the regularity
problem for parametric minimal surfaces and sketch some of the underlying ideas
involved.

Let us begin by recalling the fundamental regularity result of Hildebrandt [1] :

Theorem 5.1:   Suppose Γ is a Jordan curve in $\mathbb{R}^n$, parametrized by a
map $\gamma \in C^{m,\alpha}(\partial B; \mathbb{R}^n), m \geq 1,\ 0 < \alpha < 1$,   which is a diffeomophism of
∂B onto Γ. Then any solution $X \in \mathcal{C}(\Gamma)$ to (1.1) - (1.3) belongs to the class
$C^{m,\alpha}(\overline{B}; \mathbb{R}^n)$. Moreover, if solutions are normalized by a three-point-condition,
the $C^{m,\alpha}$−norms of solutions $X \in \mathcal{C}^*(\Gamma)$ to (1.1) - (1.3) are uniformly a-priori
bounded.

Hildebrandt originally required $m \geq 4$; the improvement to $m \geq 1$ is due
to J.C.C. Nitsche [2] . An overview of the different *proofs* of the result is given in
Nitsche [1.p. 283 ff.] Hildebrandt's approach is rather interesting in as much as it
reveals the complexity hidden in the seemingly harmless equations (1.1) - (1.3). His
basic idea is to reduce the boundary regularity problem for (1.1) - (1.3) to an interior
regularity problem for an elliptic system by means of the following transformation:

Suppose $\gamma \in C^{m,\alpha},\ m \geq 2$. Let $\gamma(w_o) = Q_o \in \Gamma$. There is a diffeomorphism
Ψ of class $C^{m,\alpha}$ of $\mathbb{R}^n$ such that Ψ maps a normal neighborhood V of
Q_o on Γ to a normal neighborhood of 0 on the (new) X^1- axis. Let

$$(5.1) \qquad\qquad\qquad Y = \Psi \circ X .$$

By harmonicity of X, Y solves an elliptic system

(5.2)
$$\Delta Y = \Gamma(Y)(\nabla Y, \nabla Y)$$

with a bounded bilinear form Γ, whose coefficients of class $C^{m-2,\alpha}$ depend continuously on Y. Γ corresponds to the Christoffel symbols of the metric

$$g_{ij}(Y) = \frac{\partial}{\partial Y^i}\Psi^{-1}(Y) \cdot \frac{\partial}{\partial Y^j}\Psi^{-1}(Y), \quad 1 \le i, j \le n.$$

By continuity, $X^{-1}(V)$ contains a neighborhood U of w_o in ∂B. The transformed surface Y thus satisfies the boundary conditions

$$Y^i = 0 \quad \text{in} \quad U, \quad i \ge 2,$$

while the conformality relations (1.2) and our choice of Ψ give a weak form of the Neumann condition
$$\partial_n Y^1 = 0 \quad \text{in } U.$$

By reflection across ∂B, the function Y hence may be extended as a solution to an elliptic system like (5.2) with *quadratic growth* in the gradient

(5.3)
$$|\Delta Y| \le a|\nabla Y|^2$$

in a full neighborhood of $w_o \in I\!R^2$. The standard interior regularity theory (cf. in particular Ladyshenskaya - Ural'ceva [1, p. 417 f.]) now enables us to bound the second derivatives of Y—and hence of X - in L^2 in a neighborhood of w_o in terms of the Dirichlet integral of X and its modulus of continuity.

By Theorem 4.9 and Proposition 4.7 both these quantities are uniformly bounded for any solution X of (1.1) -(1.3) which is normalized by a three-point-condition.

In view of Sobolev's embedding theorem an $H^{2,2}$-bound for X implies a bound for ∇X in L^p, $\forall\, p < \infty$. Returning to (5.3) the Calderon-Zygmund inequality yields that $X \in H^{2,p}$, $\forall\, p < \infty$. In particular, $\nabla X \in C^\alpha$, $\forall\, \alpha < 1$.

The complete regularity now is a consequence of Schauder's estimates for elliptic equations (5.2), cf. e.g. Gilbarg - Trudinger [1, Theorem 6.30].

☐

In later chapters we will return to this aspect and actually see some of the techniques of elliptic regularity theory in performance, cp. Section II.5.

Now we direct our attention to the regularity of the parametrized surface. Note that by the conformality relations (1.2) any solution X of (1.1)-(1.3) will be immersed in a neighborhood of points $w \in B$ where $\nabla X(w) \ne 0$.

Definition 5.2: A point $w \in \overline{B}$ is called a *branch point* of X iff $\nabla X(w) = 0$.

The behavior of X near a branch point can be analyzed by means of the following representation.

Recall that if X is harmonic, the components of the function F of $w = u + iv$ given by

$$(5.4') \qquad\qquad F = (X_u - iX_v) = \partial X$$

are holomorphic over B. Conversely, X may be reconstructed from F by complex integration

$$(5.4'') \qquad\qquad X(w) = X(w_o) + Real\left(\int_{w_o}^{w} F(w)\, dw\right).$$

Moreover, conformality is equivalent to the relation

$$(5.5) \qquad F \cdot F = 0 \qquad (\text{ componentwise complex multiplication}).$$

An interior branch point now may be characterized as a zero of the holomorphic vector function F. Since zeros of holomorphic functions are isolated this is also true for interior branch points of minimal surfaces X. Moreover, if X can be analytically extended across a segment C of ∂B X can have at most finitely many branch points on any compact subset of $B \cup C$. This observation leads to the following result of Douglas [1] and Radó [1]:

Theorem 5.3: If $X \in \mathcal{C}(\Gamma)$ is a minimal surface bounded by a Jordan arc Γ then $X|_{\partial B} : \partial B \to \Gamma$ is a homeomorphism.

Proof: It suffices to show that $X|_{\partial B}$ is injective. Assume by contradiction that $X(w_1) = X(w_2)$ for $w_1 \neq w_2 \in \partial B$. Since X maps ∂B monotonically onto Γ it follows that $X(w) \equiv X(w_1)$ for $w \in C$, where $C \subset \partial B$ is an open segment with end-points w_1, w_2. We may assume $X(w_1) = 0$. Extending X by odd reflection across C we obtain a surface

$$\hat{X}(w) = \begin{cases} X(w), & w \in B \\ -X\left(\frac{w}{|w|^2}\right), & w \notin B \end{cases}$$

which is harmonic in a neighborhood $\mathcal{N}$ of C giving rise to a holomorphic function

$$\hat{F}(w) = \left(\hat{X}_u - i\hat{X}_v\right) = \partial \hat{X}_w.$$

Moreover, $\hat{F} = F$ on B so that also $\hat{X}$ is conformal on $\mathcal{N}$. But $\hat{X} \equiv 0$ on C so that $\hat{F} \equiv 0$ on C and $\hat{F}$ must vanish identically in $\mathcal{N}$. Hence also $\hat{X} \equiv \text{const.} = 0$. In particular, $X \equiv 0$ and $X \notin \mathcal{C}(\Gamma)$. The contradiction proves the claim.

□

Definition 5.4: The *order* of a branch point w of a surface X is its order as a zero of F.

Let $w_o \in B$ be a zero of F of m–th order. Then after a rotation of coordinates

$$F(w) = a(w - w_o)^m + 0(|w - w_o|^{m+1})$$

where $a = (a^1, ... a^n) \in \mathbb{C}^n$ satisfies:

$$\mathbb{R} \ni a^1 = ia^2 > 0, \ a^3 = ... = a^n = 0,$$

as a consequence of (5.5).

Hence if $X(w_0) = 0$, X has the expansion

(5.6)
$$(X^1 + iX^2)(w) = c(w - w_0)^{m+1} + 0(|w - w_0|^{m+2})$$
$$X^j(w) = 0(|w - w_0|^{m+2}), \ j \geq 3,$$

in power series of $w - w_0$. An analoguous formula of course holds for $w_o \in \partial B$, if X is analytic in a neighborhood of w_o in $\overline{B}$. Using results of Hartman-Winter [1] it is possible to give similar expansions for X near branch points on ∂B in general provided Γ is of class C^2 or $C^{1,1}$, cp. Nitsche [1, §381] for references.

As a particular consequence of (5.6) we immediately deduce the following

Theorem 5.5:   Suppose Γ is a Jordan curve of class $C^{1,1}$, and let $X \in \mathcal{C}(\Gamma)$ be a solution to Plateau's problem (1.1) - (1.3). Then X has at most finitely many branch points. Moreover, the tangent plane to the surface X behaves continuously near any branch point.

Let us now specialize formula (5.6) to the case $n = 3$. There exist numbers $a \in \mathbb{R}, \ a > 0 , b \in \mathbb{C}$, $b \neq 0, \ l \geq 2$ such that in powers of $w - w_o$:

$$(X^1 + iX^2)(w) = a(w - w_o)^{m+1} + 0(|w - w_o|^{m+2})$$
$$X^3(w) = Real \ (b(w - w_o)^{m+l}) + 0(|w - w_o|^{m+l+1}).$$

I.e. locally, X looks like an $(m + 1)$-sheeted surface over its tangent plane through $X(w_o)$. These sheets need not all be distinct, e.g if the power series expansions for $X^1, ..., X^3$ only contain powers of $(w - w_o)^k$ for some $k \geq 2$.

Definition 5.6:   A branch point w_o of a surface X is called a *false* branch point if there exists a neighborhood U of w_o and a conformal mapping $g : U \to B, \ g(w_o) = w_o, \ g \neq \{ \text{id} \}$, such that $X \circ g = X$ near w_o.

Otherwise w_o is called a *true* branch point of X.

The following result of Gulliver, Osserman and Royden [1] - cp. also Steffen-Wente [1, Theorem 7.2] - excludes false (interior) branch points for minimal surfaces satisfying the Plateau boundary condition:

Theorem 5.7:   Suppose Γ is a rectificiable Jordan curve in $\mathbb{R}^n, \ n \geq 3$. Then a minimal surface $X \in \mathcal{C}(\Gamma)$ cannot have false interior branch points.

This result makes crucial use of Theorem 5.3.

For solutions to (1.1) - (1.3) of least area in $\mathbb{R}^3$ also true branch points can be excluded by means of the following argument due to Osserman [1].His results were completed and extended by Alt [1], Gulliver [1], Gulliver-Lesley [1] .

Theorem 5.8:   Suppose $X \in C(\Gamma)$ minimizes D in $C(\Gamma)$. Then X does not have true interior branch points. If in addition Γ is analytic then X does not have true boundary branch points, either.

The *proof* uses the fact that near interior branch points w_o by (5.6) different sheets of X must meet transversally along a branch line through w_o; cp. Chen [1] . This allows to construct a comparison surface of less area by a cutting-and-pasting-and-smoothing argument. Suppose for simplicity that x has a branch point at w_o, and let $\gamma = \gamma_1 \cup \gamma_2$:

$$\gamma_1(t) = (0, \epsilon t), \ \gamma_2(t) = (0, -\epsilon t), \ 0 < t \leq 1,$$

be a branch line of X along which $X(\gamma_1(t)) = X(\gamma_2(t))$ while the u-derivatives of X along γ_1, γ_2 are transverse. The figure illustrates the two - stage process of transforming X into a surface Y with the same area as X by a discontinuous transformation of the parameter space. (The successive images of γ_1, γ_2 are indicated for clarity.)

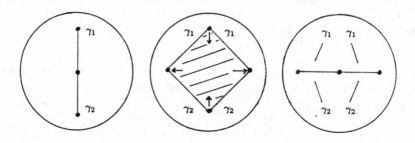

After this change of parametrization has been carried out the reparametrized surface Y will still belong to the class $C(\Gamma)$. Instead of a branch line, however, where two sheets intersect the new surface Y will have a contact line where two sheets touch. Moreover each sheet contains an edge along the contact line. Area can hence be reduced by smoothing off the edges. In this way we obtain a comparison surface $Z \in C(\Gamma)$ such that

$$\mathcal{A}(Z) < \mathcal{A}(Y) = \mathcal{A}(X) = D(X).$$

On the other hand, by (2.5) and since X by assumption minimizes D in $C(\Gamma)$

$$D(X) = \inf_{X' \in C(\Gamma)} D(X') = \inf_{X' \in C(\Gamma)} \mathcal{A}(X').$$

A contradiction.

For branch points on *analytic* boundaries the same reasoning applies, cf. Gulliver - Lesley [1]. The essential ingredient again is the existence of a branch line through the branch point.

For *smooth* wires R. Gulliver [2] has recently presented an example of a smooth wire spanning a minimal surface with a boundary branch point which is *not* connected to a branch line of the surface.

Finally, let us mention a result of Sasaki [1] and Nitsche [1, §380 , formula (156)] which relates the number and order of the branch points of a minimal surface $X \in \mathbb{R}^3$ to the total curvature of its boundary Γ by the Gauss-Bonnet formula and hence permits to estimate the former independently of X.

Theorem 5.9:   Let Γ be a Jordan curve in $\mathbb{R}^3$ of class C^2 with total curvature $\kappa(\Gamma)$. Suppose $X \in \mathcal{C}(\Gamma)$ solves (1.1) -(1.3) and has interior branch points w_j, of orders λ_j, $1 \le j \le p$, and boundary branch points $e^{i\phi_k}$ of orders ν_k, $1 \le k \le q$. Let K denote the Gaussian curvature of X. Then there holds the relation

$$1 + \sum_{j=1}^{p} \lambda_j + \sum_{k=1}^{q} \nu_k + \frac{1}{2\pi} \int_X |K| \, do \le \frac{1}{2\pi} \kappa(\Gamma).$$

In particular, if $\kappa(\Gamma) \le 4\pi$ any minimal surface spanning Γ is immersed.

Remark: Note that in case $\kappa(\Gamma) = 4\pi$ a branched minimal surface X would have to satisfy $K \equiv 0$, i.e. be a planar surface. Hence X could not have a branch point by the Riemann mapping theorem.

The following example from Nitsche [1, §288] illustrates that in general even area-minimizing parametric solutions to the Plateau problem (1.1) - (1.3) may fail to be embedded (and hence will be physically unstable):

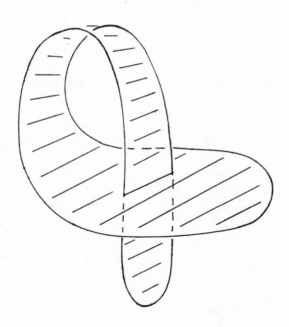

For curves like the depicted one the true physical solutions apparently can be described best by the methods of geometric measure theory, cf. Almgren [1].

However, there is a class of curves in $I\!R^3$ where the least-area solution to the parametric Plateau problem can be shown to be a minimal embedding: these are the so-called extreme curves, i.e. Jordan curves on the boundaries of convex regions $\Omega \in I\!R^3$. More generally, one can also allow curves on boundaries of regions Ω with the property that the mean curvarure of $\partial\Omega$ with respect to the interior normal is non-negative ("$M-$convex" regions).

Convex or M-convex surfaces provide natural "barriers" for minimal surfaces (by the maximum principle for the non-parametric minimal surface eqution, cf. Nitsche [1, §579 ff.]). The following result is due to Meeks and Yau[1]; the existence of an embedded minimal disc was established independently by Tomi and Tromba [1] , resp. Almgren and Simon [1] :

__Theorem 5.10:__   Let Ω be an M-convex region in $I\!R^3$ of class C^2 and let $\Gamma \subset \partial\Omega$ be a rectifiable Jordan curve, which is contractible in $\overline{\Omega}$. Then there exists an embedded minimal disc with boundary Γ. Moreover, any solution X of (1.1)-(1.3) with $X(B) \subset \Omega$ which minimizes D in this class is embedded.

As a special case Theorem 5.10 contains the "existence" part of the following classical result of Radó [2]. The uniqueness is a consequence of the maximum principle.

__Theorem 5.11:__   Suppose Γ is a Jordan curve in $I\!R^3$ having a single-valued parallel projection onto a convex curve $\tilde{\Gamma}$ in some plane P in $I\!R^3$. Then there exist a unique minimal surface spanning Γ (up to conformal reparametrization). This surface is a graph over the region bounded by $\tilde{\Gamma}$ in P.

In higher dimensions a result like Theorem 5.10 is not known.

Appendix

We establish (2.5). For simplicity we assume in addition that $\Gamma \in C^2$.

Proposition A.1:   Let Γ be a $C^2 -$ Jordan curve. Then $\mathcal{C}(\Gamma) \neq \emptyset$, and

$$\inf_{X \in \mathcal{C}(\Gamma)} D(X) = \inf_{X \in \mathcal{C}(\Gamma)} A(X).$$

Proof: Let $X \in \mathcal{C}(\Gamma) \cap C^2(\overline{B}; \mathbb{R}^n)$. Approximate X by embedded surfaces $X_\epsilon(u, v) \equiv (X(u, v), \epsilon u, \epsilon v) \in C^2(\overline{B}; \mathbb{R}^{n+2})$. We claim that for any $\epsilon > 0$ there exists a map $g \in H^{1,2} \cap C^0(\overline{B}; \mathbb{R}^2)$ of $\overline{B}$ onto itself mapping ∂B monotonically onto ∂B and such that

(A.1) $D(X_\epsilon \circ g) = A(X_\epsilon)$.

Since for any $\epsilon > 0$, and any such g there holds

$$X \circ g \in \mathcal{C}(\Gamma), \ D(X \circ g) \leq D(X_\epsilon \circ g),$$

while as $\epsilon \to 0$ clearly $A(X_\epsilon) \to A(X)$, we infer from (A.1) that

$$\inf_{X \in \mathcal{C}(\Gamma)} D(X) \leq \inf_{X \in \mathcal{C}(\Gamma) \cap C^2(\overline{B}; \mathbf{R}^n)} A(X).$$

By density of $C^2(\overline{B}, \mathbb{R}^n)$ in $\mathcal{C}(\Gamma)$ and continuity of A the latter inequality implies the Proposition.

In order to prove (A.1) introduce the set $\mathcal{F}$ as the weak closure in $H^{1,2}(B; \mathbb{R}^2)$ of the set

$$\mathcal{F} = \Big\{ g \in C^1(\overline{B}; \mathbb{R}^2) \ | $$
$$g \text{ is a diffeomorphism onto } \overline{B}, \ g\left(e^{\frac{2\pi i k}{3}}\right) = e^{\frac{2\pi i k}{3}}, \ k = 1, 2, 3 \Big\}$$

of normalized diffeomorphisms of $\overline{B}$.

Note

Lemma A.2: $\mathcal{F}$ is weakly closed in $H^{1,2}(B; \mathbb{R}^2)$. For any $g \in \mathcal{F}$, any $Z \in C^*(\Gamma) \cap C^2(\overline{B}; \mathbb{R}^n)$ we have $Z \circ g \in C^*(\Gamma)$, $A(Z \circ g) = A(Z)$.

Proof: If $g_m^k \overset{w}{\rightharpoonup} g^k \ (m \to \infty)$, $g^k \overset{w}{\rightharpoonup} g(k \to \infty)$ weakly in $H^{1,2}(B; \mathbb{R}^2)$

where $g_m^k \in \mathcal{F}$, since $H^{1,2}(B; \mathbb{R}^2)$ is separable, a diagonal sequence $g_{m(k)}^k \overset{w}{\rightharpoonup} g$, and $\mathcal{F}$ is weakly closed.

Next we show that any $g \in \overline{\mathcal{F}}$ is in fact a uniform limit of diffeomorphisms. Indeed, a standard argument based on the Courant - Lebesgue Lemma 4.4 shows that $H^{1,2}-$bounded subsets of $\mathcal{F}$ are equi-continuous:

First, remark that $\mathcal{F} \subset C^*(\partial B)$, whence bounded subsets of $\mathcal{F}$ are equi-continuous on ∂B, cp. Lemma 4.3.

Next, given $K > 0, \mu > 0$ there exists $\nu > 0$ such that for any $g \in \mathcal{F}$ with $\mathcal{F}(g) \leq K$, any $w_o \in \overline{B}$ there is a radius $\rho \in [\nu, \sqrt{\nu}]$ such that on $C_\rho = \partial B_\rho(w_o) \cap B$:

$$\sup_{w,w' \in C_\rho} |g(w) - g(w')|^2 \leq 2\pi\rho \int_{C_\rho} |\partial_s g|^2 ds \leq c \frac{D(g)}{|\ln \nu|} < \mu^2.$$

Since g is a diffeomorphism and since the set $\{g \in \mathcal{F} \mid D(g) \leq K\}$ is equi-continuous on ∂B, for small $\mu > 0$ any such g maps the disc $B_\rho(w_o) \cap B$ onto the "small" disc bounded by $g(C_\rho)$. Hence

$$\sup_{|w-w'|<\nu} |g(w) - g(w')| \leq \sup_{w_o \in \overline{B}} \left\{ \sup_{w,w' \in B_\rho(w_o) \cap B} |g(w) - g(w')| \right\}.$$

$$\leq \sup_{w_o \in \overline{B}} \left\{ \sup_{w,w' \in C_\rho} |g(w) - g(w')| \right\} < \mu.$$

This proves equi-continuity of $H^{1,2}$-bounded subsets of $\mathcal{F}$.

In particular, any $g \in \overline{\mathcal{F}}$ is a continuous map of $\overline{B}$ onto $\overline{B}$ mapping ∂B monotonically onto ∂B and satisfying the three-point-condition $g\left(e^{\frac{2\pi i k}{3}}\right) = e^{\frac{2\pi i k}{3}}$, $k = 1, 2, 3$. Thus, for $Z \in C^*(\Gamma) \cap C^2(\overline{B}; \mathbb{R}^n)$ also $Z \circ g \in C^*(\Gamma)$. Moreover, if $g_m \overset{w}{\to} g$ weakly in $H^{1,2}(B; \mathbb{R}^2)$, $g_m \in \mathcal{F}$, we have

(A.2)
$$\det(\nabla g_m) = g^1_{m_u} g^2_{m_v} - g^1_{m_v} g^2_{m_u} =$$
$$= (g^1_m g^2_{m_v})_u - (g^1_m g^2_{m_u})_v \overset{w}{\to} (g^1 g^2_v)_u - (g^1 g^2_u)_v = \det(\nabla g)$$

in the sense of distributions. We use this fact to compute $A(Z \circ g)$. First, since $g_m \to g$ uniformly there holds

$$A(Z) = A(Z \circ g_m) = \int_B \sqrt{\det((\nabla Z \cdot \nabla Z) \circ g_m)} \det(\nabla g_m) \, dw$$

$$= \int_B \sqrt{\det((\nabla Z \cdot \nabla Z) \circ g)} \det(\nabla g_m) \, dw + o(1)$$

with $o(1) \to 0$. By (A.2) this equals

$$= -\int_B \left(g^1_m g^2_{m_v} \partial_u - g^1_m g^2_{m_u} \partial_v\right) \sqrt{\det((\nabla Z \cdot \nabla Z) \circ g)} \, dw$$

$$+ \int_{\partial B} g^1_m \partial_\tau g^2_m \sqrt{\det((\nabla Z \cdot \nabla Z) \circ g)} \, do + o(1),$$

∂_τ denoting the tangent derivative in counter clock-wise direction. Since g_m maps ∂B monotonically onto ∂B we have $\partial_\tau g_m \in L^1$ with $\int_{\partial B} |\partial_\tau g_m| do = 2\pi$. Hence we may pass to the limit $m \to \infty$ in both integrals and finally arrive at the identity

$$A(Z \circ g) = \lim_{m \to \infty} A(Z \circ g_m) = A(Z),$$

as claimed.

□

Proof of Proposition A.1 (completed): Now consider the functional E on $\overline{\mathcal{F}}$:

$$E(g) = D(X_\epsilon \circ g) = \frac{1}{2} \int_B |(\nabla X_\epsilon) \circ g) \cdot \nabla g|^2 dw.$$

By choice of X_ϵ we have

$$E(g) \geq \epsilon D(g)$$

and by Example 3.5 E is coercive on $\overline{\mathcal{F}}$ with respect to the $H^{1,2}$-topology.

Let $g_m \in \overline{\mathcal{F}}$ be a minimizing sequence such that

$$E(g_m) \to \inf_{g \in \overline{\mathcal{F}}} E(g).$$

By Proposition 4.5 we may assume that $g_m \xrightarrow{w} g \in \overline{\mathcal{F}}$ and $X_\epsilon \circ g_m \xrightarrow{w} X_\epsilon \circ g \in C^*(\Gamma)$, whence by weak lower semi-continuity of Dirichlet's integral

$$D(X_\epsilon \circ g) = E(g) \leq \liminf_{m \to \infty} D(X_\epsilon \circ g_m) = \liminf_{m \to \infty} E(g_m).$$

I.e. $X_\epsilon \circ g$ minimizes D among all surfaces $X_\epsilon \circ g'$, $g' \in \overline{\mathcal{F}}$. In particular, $X_\epsilon \circ g$ satisfies the hypothesis of Lemma 2.4 and is conformal. But then by Lemma A.2

$$A(X_\epsilon) = A(X_\epsilon \circ g) = D(X_\epsilon \circ g) = \inf_{g' \in \overline{\mathcal{F}}} D(X_\epsilon \circ g'),$$

proving (A.1) and the Proposition.

□

II. Unstable minimal surfaces

1. Ljusternik-Schnirelman theory on convex sets in Banach spaces.

The method of gradient line deformations and the minimax-principle are the most general avaible tools for obtaining unstable critical points in the calculus of variations. Historically, the use of these methods can be traced back to the beginning of this century, cf. Birkhoff's [1] theorem on the existence of closed geodesics on surfaces of genus 0 . Through their famous improvement of Birkhoff's result the names of Ljusternik and Schnirelman [1] became intimately attached to these methods. In 1964 a major extension of these techniques was proposed by Palais [1], [2], Smale [1] and Palais - Smale [1]. Their fundamental work has found many applications and has inspired a lot of further research. One of the most significant contributions may be the well-known paper of A. Ambrosetti and P. H. Rabinowitz [1].

For the Plateau Problem, however, these methods still seemed inadequate, and for a long time analysts believed that the Palais-Smale condition would "never" be satisfied in any "hard" variational problem as the Plateau problem for minimal surfaces was considered to be. (Cf. the remarks by Hildebrandt [4, p. 323 f.])

As we shall see, by a simple variation of the classical concepts of Ljusternik-Schnirelman, resp. Palais and Smale the Plateau problem can be naturally incorporated in the frame of these methods. This extension of Ljusternik-Schnirelmann theory and its application to the Plateau problem was presented in Struwe [1]. In abstract terms we may regard this method as an extension of Ljusternik-Schnirelman theory to functionals defined on closed convex sets of (affine) Banach spaces and satisfying a variant of the Palais - Smale condition.

We now recall the pertinent ideas. Throughout this section we make the following

Assumption:

Let T be a Banach space with norm $|\cdot|$, $M \subset T$ closed and convex. Suppose $E : M \to I\!R$ admits a Frechet differentiable extension $E \in C^1(T; I\!R)$.

Definition 1.1: At a point $x \in M$ let

$$g(x) = \sup_{\substack{y \in M \\ |x-y| < 1}} \langle dE(x),\ x - y \rangle$$

measure the *slope* of E *in* M.

Remark 1.2.   i) If $M = T$, then $g(x) = |dE(x)|$.

ii) Since $E \in C^1$ and M is convex , g is continuous.

Definition 1.3:   A point $x \in M$ is called *critical* iff $g(x) = 0$. Otherwise x is called *regular*. If x is critical, $\beta = E(x)$ is called a *critical value* of E. If $E^{-1}(\beta)$ consists only of regular points, β is called *regular*.

Remark 1.4:   If $M = T$ by Remark 1.2 Definition 1.3 coincides with the usual definition of critical points.

Definition 1.5:   Let $\tilde{M} = \{x \in M | g(x) \neq 0\}$. A Lipschitz continuous mapping $\tilde{e} : \tilde{M} \to T$ is a *pseudo gradient vector field for* E *on* M if the following holds:

i) $\tilde{e}(x) + x \in M, \quad \forall\ x \in M,$

ii)   There exists a constant $c > 0$ such that
   a) $|\tilde{e}(x)| < 1,$
   b) $\langle dE(x),\ \tilde{e}(x) \rangle < -\min\{c^{-1}g(x)^2,\ 1\}.$

Exactly as in the case $M = T$, cf. Palais [3, Chapter 3], we now establish

Lemma 1.6:   There exists a pseudo-gradient vector field $\tilde{e}$ for E on M.

Proof: Let $x_0 \in \tilde{M}$ and choose $y_0 \in M$ such that

$$|x_0 - y_0| < \min\{g(x_0),\ 1\},$$
(1.1)
$$\langle dE(x_0),\ x_0 - y_0 \rangle > \min\{1/2\ g(x_0)^2,\ 1\}.$$

By continuity, there exists a neighborhood $V(x_0)$ of x_0 in M such that for $x \in V(x_0)$ we still have $|x - y_0| < \min\{g(x),\ 1\}$ while $\langle dE(x),\ x - y_0 \rangle > \min\{1/2\ g(x)^2,\ 1\}$. Hence $e_0(x) := y_0 - x$ is a pseudo-gradient vector field for E on $V(x_0)$.

The sets $\{V(x_0)_{x_0 \in \tilde{M}}$ are an open cover of $\tilde{M}$. Since $\tilde{M} \subset T$ is metric there exists a locally finite refinement $\{V(x_\iota)\}_{\iota \in I}$ of this cover, i.e. having the property that for any $x_0 \in M$ there is a neighborhood V_0 of x_0 and a finite collection $I_0 \subset I$ such that for $\iota \in I \backslash I_0$ we have $V_o \cap V(x_\iota) = \emptyset$, cf. Kelly [1, Thm. 8, p.156; Cor. 35, p. 160]. Let $\{\psi_\iota\}_{\iota \in I}$ be a Lipschitz continuous partition of unity subordinate to $\{V(x_\iota)\}_{\iota \in I}$, i.e a collection of Lipschitz continuous functions ψ_ι with support in $V(x_\iota)$ such that $0 \leq \psi_\iota \leq 1$ for each $\iota \in I$ and

$$\sum_{\iota' \in I} \psi_{\iota'}(x) = 1, \quad \forall\ x \in \tilde{M}.$$

E. g. we may let

$$\rho_\iota(x) = \inf_{y \in M \setminus V(x_\iota)} |x - y|$$

$$\psi_\iota(x) = \frac{\rho_\iota(x)}{\displaystyle\sum_{\iota' \in I} \rho_{\iota'}(x)}.$$

Finally, define

$$\tilde{e}(x) = \sum_{\iota \in I} \psi_\iota(x)(y_\iota - x),$$

where $y_\iota \in M$ is associated to x_ι by (1.1). $\tilde{e}$ is Lipschitz continuous and satisfies i) and ii) of Definition 1.7

□

Definition 1.7: E satisfies the *Palais-Smale condition on* M if the following holds:

(P.S.)   Any sequence $\{x_m\}$ in M such that $|E(x_m)| \le c$ uniformly, while $g(x_m) \to 0$ $(m \to \infty)$ is relatively compact.

Remark 1.8:   Again (P.S.) reduces to a variant of the well-known Palais-Smale condition (C), cf. Palais - Smale [1], in case $M = T$.

The Palais-Smale condition crucially enters in the following fundamental "deformation lemma."
For $\beta \in I\!R$ let

$$M_\beta = \{x \in M \mid E(x) < \beta\},$$

$$K_\beta = \{x \in M \mid E(x) = \beta,\ g(x) = 0\}.$$

Lemma 1.9:   Suppose E satisfies (P.S.) on M. Let $\beta \in I\!R$, $\bar{\epsilon} > 0$, and suppose N is a neighborhood of K_β in M.
Then there exists a number $\epsilon \in]0, \bar{\epsilon}[$ and a continuous one-parameter family $\Phi : [0,1] \times M \to M$ of continuous maps $\Phi(t, \cdot)$ of M having the properties

i) $\Phi(t,x) = x$ if $t = 0$, or if $|E(x) - \beta| \ge \bar{\epsilon}$, or if $g(x) = 0$.

ii) $E(\Phi(t,x))$ is non-increasing in t for any x,

iii) $\Phi(1, M_{\beta+\epsilon} \setminus N) \subset M_{\beta-\epsilon}$,
$\Phi(1, M_{\beta+\epsilon}) \subset M_{\beta-\epsilon} \cup N$.

For the proof we need the following auxiliary

Lemma 1.10:   Suppose E satisfies (P.S.) on M. Then for any $\beta \in I\!R$ the families

$$N_{\beta,\delta} = \{x \in M \mid |E(x) - \beta| < \delta,\ g(x) < \delta\},\ \delta > 0$$

resp.

$$U_{\beta,\rho} = \{x \in M \mid |x - y| < \rho \text{ for some } y \in K_\beta\}, \rho > 0$$

constitute fundamental systems of neighborhoods of K_β.

Proof: By continuity of g, clearly each $N_{\beta,\delta}$ and each $U_{\beta,\rho}$ is a neighborhood of K_β. Hence it remains to show that any neighborhood N of K_β contains at least one of the sets $N_{\beta,\delta}, U_{\beta,\rho}$.

Suppose by contradiction that for some neighborhood N of K_β and any $\delta > 0$ we have $N_{\beta,\delta} \not\subset N$. Then for a sequence $\delta_m \to 0$ there exist elements $x_m \in N_{\beta,\delta_m} \setminus N$. By (P.S.) the sequence $\{x_m\}$ accumulates at a critical point $x \in K_\beta$. Hence $x_m \in N$ for large m, a contradiction. Similarly, if $U_{\beta,\rho_m} \not\subset N$ for $\rho_m \to 0$ there exist sequences $x_m \in U_{\beta,\rho_m} \setminus N$, $y_m \in K_\beta$ such that $|x_m - y_m| \leq 2\rho_m$. By (P.S.) $\{y_m\}$ accumulates at $y \in K_\beta$, hence also $\{x_m\}$ does. A contradiction.

☐

Proof of Lemma 1.9: Coose numbers $0 < \delta < \delta' \leq 1$, $0 < \rho \leq 1$ such that

$$N \supset U_{\beta,\rho} \supset U_{\beta,\frac{\rho}{2}} \supset N_{\beta,\delta'} \supset N_{\beta,\delta}$$

and let η be a Lipschitz continuous function on M such that $0 \leq \eta \leq 1$ and $\eta \equiv 0$ in $N_{\beta,\delta}$, $\eta \equiv 1$ outside $N_{\beta,\delta'}$. Also let $\epsilon < 1/2 \min\{\delta, \bar{\epsilon}\}$ be a number to be specified later, and choose a function $\varphi \in C_0^\infty(I\!R)$ such that $0 \leq \varphi \leq 1$, $\varphi(s) = 1$ if $|s - \beta| < \epsilon$, $\varphi(s) = 0$ if $|s - \beta| \geq 2\epsilon$.

Define

$$e(x) = \begin{cases} \varphi(E(x)) \, \eta(x) \tilde{e}(x), & x \in \tilde{M} \\ 0 & , \text{ else,} \end{cases}$$

where $\tilde{e}$ is the pseudo-gradient vector field constructed in Lemma 1.8. Since any critical point of E on M has a neighborhood where either φ or η vanishes, e is Lipschitz continuous. Moreover, by convexity of M, e satisfies

(1.2) $$e(x) + x \in M$$

at any point $x \in M$.

Now let $\Phi : [0, \infty[\times M \to M$ be the solution to the initial value problem

(1.3)
$$\frac{\partial}{\partial t}\Phi(t, x) = e(\Phi(t, x)),$$
$$\Phi(0, x) = x.$$

By convexity of M, Φ may be constructed as a limit of approximate trajectories of (1.3) by Euler's method, cp. Struwe [1, Lemma 3.8].

If T is locally strictly convex and if the projection $p_m : T \to M$ of T onto M obtained by letting

$$|p_M(x) - x| = \inf_{y \in M} |y - x|$$

is locally Lipschitz continuous (e.g. if T is a Hilbert space), then a more instructive existence proof goes as follows: Extend e to T by letting

$$e(x) = e(p_M(x)) .$$

Now let $\overline{\Phi} : [0, \infty[\times T \to T$ be the solution to (1.3) on T which exists globally by Lipschitz continuity and boundedness of e. By (1.2) M is an *invariant region* for $\overline{\Phi}$ and the deformation Φ may be obtained by restricting $\overline{\Phi}$ to M.

As Φ solves (1.3), each $\Phi(t, \cdot)$ now is a continuous map from M into M and Φ trivially satisfies i), ii) by our choice of ϵ, η. Finally, if $E(x) \leq \beta + \epsilon$, either $E(\Phi(1, x)) \leq \beta - \epsilon$ or $|E(\Phi(t, x)) - \beta| \leq \epsilon$ for all $t \in [o, 1]$. In the latter case, moreover, by choice of η :

$$E(\Phi(1, x)) = E(x) + \int_0^1 \frac{d}{dt} E(\Phi(t, x)) dt$$

$$\text{(1.4)} \qquad \leq \beta + \epsilon + \int_0^1 \eta(\Phi(x, t)) \langle dE(\cdot), \, \tilde{e}(\cdot) \rangle dt$$

$$\leq \beta + \epsilon - \int_0^1 \eta(\cdot) \, \min\{1/2 \, g(\cdot)^2, 1\} dt$$

$$\leq \beta + \epsilon - 1/2 \, \delta^2 \cdot |\{t \in [0, 1] | \Phi(t, x) \notin N_{\beta, \delta'}\}|.$$

Now suppose $x \notin N$ or $\Phi(1, x) \notin N$. Then either $\Phi(t, x) \notin N_{\beta, \delta'}$ for all $t \in [0, 1]$, or the flow $\Phi(t, x)$ through x must traverse the annulus $U_{\beta, \rho} \backslash U_{\beta, \rho/2}$. By boundedness $|e(x)| \leq 1$ this will require

$$|\{t \in [0, 1] \mid \Phi(t, x) \notin N_{\beta, \delta'}\}| \geq \rho/2,$$

and hence in any event we obtain that

$$E(\Phi(t, x)) \leq \beta + \epsilon - \frac{1}{4} \rho \delta^2.$$

Thus iii) will be satisfied if we let

$$\epsilon = 1/2 \, \min\{\overline{\epsilon}, \frac{1}{4} \rho \delta^2\}.$$

This completes the construction.

$$\square$$

A variant of the deformation lemma 1.9 yields the following result:

Lemma 1.11:   Suppose E satisfies (P.S.) on M, and let $x_0 \in M$ be a strict relative minimum of E in M. Then there exists a number $\epsilon_0 > 0$ such that for any $\epsilon \in]0, \epsilon_0[$ we have

$$\inf_{\substack{x \in M \\ |x - x_0| = \epsilon}} E(x) > E(x_0).$$

Proof: By assumption there exists $\epsilon_0 > 0$ such that

$$E(x) > E(x_0), \quad \forall x \in M, \ 0 < |x - x_0| < 2\epsilon_0.$$

Choose $\epsilon \in]0, \epsilon_0[$ and let $\{x_m\}$ be a minimizing sequence for E in $S_\epsilon(x_o) := \{x \in M | \ |x - x_0| = \epsilon\}$,

$$E(x_m) \to \inf_{x \in S_\epsilon(x_0)} E(x) =: \beta.$$

If $\beta > E(x_o)$ the proof is complete. Otherwise $E(x_m) \to E(x_0)$ and either $g(x_m) \to 0$ or there exists $\delta_0 > 0$ such that $g(x_m) \geq \delta_0$ for all m.

In the first case, by (P.S.) $\{x_m\}$ accumulates at an element $x \in S_\epsilon(x_0)$, where $E(x) = E(x_0)$. Since this contradicts the strict minimality of x_0, we are left with the second case.

Choose $N = N_{\beta, \delta_0} \supset U_{\beta, \rho} \supset U_{\beta, \rho/2} \supset N_{\beta, \delta'} \supset N_{\beta, \delta}$ and let φ, η, e, Φ be defined as in Lemma 1.9. Consider the sequence $y_m := \Phi(\frac{\epsilon}{2}, x_m)$. Since $|e| \leq 1$ it follows that

$$0 < \frac{\epsilon}{2} \leq |y_m - x_0| \leq \frac{3\epsilon}{2} < 2\epsilon_0.$$

Moreover, since $\beta = E(x_0) \leq E(\Phi(t, x_m)) \leq E(x_m) \leq \beta + \epsilon$ for large m and $t \leq \frac{\epsilon}{2}$, like (1.4) we obtain

$$E(y_m) = E(x_m) - \frac{1}{2} \delta^2 |\{t \in [0, \frac{\epsilon}{2}]| \ \Phi(t, x) \notin N_{\beta, \delta'}\}|$$

$$\leq E(x_m) - \frac{1}{2} \delta^2 \cdot \min\{\frac{\epsilon}{2}, \frac{\rho}{2}\}.$$

But since $E(x_m) \to E(x_0) \ (m \to \infty)$ this implies that $E(y_m) < E(x_0)$ for large m. The contradiction proves the lemma.

$$\square$$

Lemmata 1.9, 1.11 immediately yield the following variant of the classical "mountain-pass-lemma":

Theorem 1.12:   Suppose E satisfies (P.S.) on M, and let x_1, x_2 be distinct strict relative minima of E. Then E possesses a third critical point x_3 distinct from x_1, x_2. x_3 is characterized by the minimax-principle

$$(1.5') \qquad\qquad E(x_3) = \inf_{p \in P} \sup_{x \in p} E(x) =: \beta$$

where

$$(1.5'') \qquad\qquad P = \{p \in C([0, 1]; \ M) \ | \ p(0) = x_1, \ p(1) = x_2\}.$$

Moreover

$$E(x_3) > \sup\{E(x_1),\ E(x_2)\},$$

and x_3 is unstable in the sense that x_3 is not a relative minimum of E.

Proof: i) By Lemma 1.11 $\beta > \sup\{E(x_1),\ E(x_2)\}$. Suppose by contradiction that β is a regular value of E, i.e. $K_\beta = 0$. Choose $N = \emptyset$, $\bar{\epsilon} = 1$ and let $\epsilon > 0, \Phi$ be as constructed in Lemma 1.9.

By definition of β there is $p \in P$ such that

$$\sup_{x \in p} E(x) < \beta + \epsilon.$$

Applying the map $\Phi(1, \cdot)$ to p by property i) of Φ the path $p' = \Phi(1, p) \in P$. while by iii)

$$\sup_{x \in p'} E(x) < \beta - \epsilon.$$

The contradiction shows that β is critical.

ii)   Now suppose that E possesses only critical points of energy β which are relative minimizers of E in M. The set K_β will then be both open and closed in $\overline{M}_\beta = \{x \in M | E(x) \le \beta\}$; hence there exists a neighborhood N of K_β in M such that N and $\overline{M}_\beta \backslash K_\beta$ are disjoint. A fortiori, then $M_{\beta - \epsilon}$ and N will be disconnected for any $\epsilon > 0$. Choosing $\epsilon > 0$, Φ corresponding to this N and $\bar{\epsilon} = 1$, and letting $p \subset M_{\beta + \epsilon}$, however by property iii) of Φ we obtain a path

$$p' = \Phi(1, p) \in P,\ p' \subset M_{\beta - \epsilon} \cup N.$$

Since $p' \ni x_1 \notin N$ and since $M_{\beta - \epsilon}$ and N are disconnected, $p' \subset M_{\beta - \epsilon}$. The contradiction shows that E has an unstable critical point of energy β.

□

A slight variant of the preceding result is given in

Theorem 1.13: Suppose E satisfies (P.S.), and let x_1, x_2 be two (not neccessarily strict) relative minima of E. Then either $E(x_1) = E(x_2) = \beta_0$ and $x_1,\ x_2$ are connected in any neighborhood of K_{β_0}, or there exists an unstable critical point x_3 of E characterized by the minimax-principle (1.5).

Proof: Let β be given by (1.5). If K_β consists only of relative minimizers of E as in part ii) of the proof of Theorem 1.12 we deduce that for any sufficiently small neighborhood N of K_β there holds $N \cap M_{\beta - \epsilon} = \emptyset$ for any $\epsilon > 0$. Letting $\epsilon > 0$, Φ be as constructed in Lemma 1.9 corresponding to $N, \bar{\epsilon} = 1$, and choosing $p \in P$ such that $p \subset M_{\beta + \epsilon}$, we obtain a path $p' = \Phi(1, p) \in P$ connecting

x_1 with x_2 in $M_{\beta-\epsilon} \cup N$. Hence $p' \subset N$ and x_1 and x_2 both belong to the same connected component of K_β, in particular $E(x_1) = E(x_2) = \beta$.

$\square$

Along the same lines numerous other existence results for unstable critical points can be given. For our purpose, however, Theorems 1.12, 1.13 will suffice and we refer the interested reader to Palais [3] or Ambrosetti - Rabinowitz [1]. Theorem 1.13 is related to a result by Pucci and Serrin [1] .

2. The mountain-pass-lemma for minimal surfaces.

In order to convey the preceding results to the Plateau problem we reformulate the variational problem in a more convenient way. At first we closely follow Douglas' original approach to the Plateau problem.

Let $\gamma : \partial B \to \Gamma$ be a reference parametrization of the Jordan curve Γ. We assume that γ is a homeomorphism.

Note that by (1.1) it suffices to consider surfaces X whose coordinate functions are harmonic:
$$\mathcal{C}_0(\Gamma) = \{X \in \mathcal{C}(\Gamma) | \triangle X = 0\}.$$

By composition with γ and harmonic extension $\mathcal{C}_0(\Gamma)$ may be represented by the space of monotone reparametrizations of $\partial B \cong I\!\!R/2\pi$.

More precisely, let $h : C^0(\partial B) \to C^0(\overline{B})$ be the harmonic extension operator defined by
$$\triangle h(\varphi) = 0 \text{ in } B, \quad h(\varphi) = \varphi \text{ on } \partial B.$$

h is linear and continuous by the maximum principle. Also let $e^{i\phi}$ denote coordinates on ∂B. Then the map

(2.1')
$$X : x \mapsto h(\gamma \circ x),$$

where $\gamma \circ x$ denotes the map

(2.1")
$$(\gamma \circ x)(e^{i\phi}) := \gamma(e^{ix(\phi)}) =: \gamma(x(\phi)),$$

is a local homeomorphism between

(2.2)
$$M = \{x \in C^0(I\!\!R; I\!\!R) \mid x \text{ monotone};$$
$$x(\phi + 2\pi) = x(\phi) + 2\pi, \forall \phi; D(X(x)) < \infty\}.$$

and $\mathcal{C}_0(\Gamma)$ with respect to the C^0-topology. Introducing a three point-condition, in a similar way the set
$$C_0^*(\Gamma) = \{X \in C^*(\Gamma) | \triangle X = 0\}$$

can be represented via (2.1) in terms of

(2.3)
$$M^* = \{x \in M | x \left(\frac{2\pi k}{3} \right) = \frac{2\pi k}{3}, \forall k \in \mathbb{Z}\}.$$

Finally, for $x \in M$ let

(2.4)
$$E(x) = D(X(x)) = \frac{1}{16\pi} \int_0^{2\pi} \int_0^{2\pi} \frac{|\gamma(x(\phi)) - \gamma(x(\phi'))|^2}{\sin^2 \left(\dfrac{\phi - \phi'}{2} \right)} d\phi \, d\phi',$$

denote the Douglas - Dirichlet integral of $X(x) = h(\gamma \circ x)$, cf. Nitsche [1; §§ 310 - 311].

Proposition I.4.7 in this notation takes the form:

Proposition 2.1:   For any $\alpha \in I\!R$ the set

$$\{x \in M^* \mid E(x) \le \alpha\}$$

is compact with respect to the C^0-topology.

For the larger set M we have the following weaker compactness result which we note for later convenience:

Propsition 2.2:   Suppose $\{x_m\}$ is a sequence in M such that $|x_m(0)| \le c$ uniformly. Suppose $E(x_m) \le c$ uniformly. Then a subsequence $x_m \to x$ in $L^1_{\text{loc}}(I\!R)$ and either

$$x(\phi) \equiv \quad \text{const} \ (\text{mod } 2\pi),$$

or

$$x_m \to x \quad \text{uniformly} \ (m \to \infty),$$

and $x \in M$.

Proof: Local L^1-compactness of $\{x_m\}$ follows from Helly's theorem and monotonicity of x_m. Note that x is monotone; hence if x is continuous then $x_m \to x$ uniformly by Dini's theorem, and $x \in M$.

In any event, letting $X_m = X(x_m)$, $X = X(x) = h(\gamma \circ x)$, by uniform boundedness of $E(x_m)$ and weak lower semi-continuity of $D : E(x) < \infty$. But finiteness of Douglas' integral (2.4) and monotonicity of x imply continuity of $\gamma \circ x$. Hence, if x is discontinuous it follows that $\gamma \circ x \equiv \text{const}$, i.e. $x \equiv \text{const}$. $(\text{mod } 2\pi)$.

$$\square$$

Note that by weak lower semi-continuity of D on $H^{1,2}(B)$ the functional E will be lower semi-continuous on M, if we endow M with the topology of C^o.

We now leave the trails of Douglas and his followers and define a new topology on M which will render E differentiable (and even smooth for smooth Γ)!

Throughout the following assume γ is a diffeomorphism of class C^r, $r \ge 2$.

Note that M is a convex subset of the affine space

$$\{ \text{id} \} + C^0(I\!R/2\pi).$$

However, M is not closed in the induced topology (if we only allow finite values of E).

Now let

(2.5) $$T = H^{1/2,2} \cap C^0(\mathbb{R}/2\pi),$$

where the seminorm in $H^{1/2,2}(\mathbb{R}/2\pi)$ is given by

(2.6) $$|x|^2_{1/2} = \int_0^{4\pi}\int_0^{4\pi} \frac{|x(\phi) - x(\phi')|^2}{|\phi - \phi'|^2} \, d\phi \, d\phi'$$

and the norm $|\cdot|$ in T is that induced by $|\cdot|_{1/2}$ and $|\cdot|_{L^\infty}$.

Lemma 2.3: M is a closed convex subset of the affine space $\{\text{id}\} + T$.

Proof: It suffices to show that finiteness $E(x) < \infty$ implies that the expression (2.6) is finite and vice verse.

But if $[s] = \inf\{|s - 2m\pi| \, | \, m \in \mathbb{Z}\}$ the integral (2.4)

$$E(x) = \frac{1}{16\pi} \int_0^{2\pi}\int_0^{2\pi} \frac{|\gamma(x(\phi)) - \gamma(x(\phi'))|^2}{\sin^2\left(\dfrac{\phi - \phi'}{2}\right)} \, d\phi \, d\phi'$$

is equivalent to the expression

$$\int_0^{2\pi}\int_0^{2\pi} \frac{|\gamma(x(\phi)) - \gamma(x(\phi'))|^2}{[x(\phi) - x(\phi')]^2} \cdot \frac{[x(\phi) - x(\phi')]^2}{[\phi - \phi']^2} \, d\phi \, d\phi'.$$

The latter in turn may be estimated from above

$$\leq \|\frac{d}{d\phi}\gamma\|^2_\infty \cdot \int_0^{2\pi}\int_0^{2\pi} \frac{[x(\phi) - x(\phi')]^2}{[\phi - \phi']^2} \, d\phi \, d\phi'$$

$$\leq \|\frac{d}{d\phi}\gamma\|^2_\infty \cdot \int_0^{4\pi}\int_0^{4\pi} \frac{|x(\phi) - x(\phi')|^2}{|\phi - \phi'|^2} \, d\phi \, d\phi'.$$

To obtain an estimate from below note that since γ is a diffeomorphism there exists a constant $c_\gamma > 0$ such that

$$\frac{|\gamma(s) - \gamma(t)|^2}{|s - t|^2} \geq c_\gamma, \quad \forall \, s,t \, : \, 0 < |s - t| < \pi.$$

Hence

$$E(x) \geq c_\gamma \int_0^{2\pi}\int_0^{2\pi} \frac{[x(\phi) - x(\phi')]^2}{[\phi - \phi']^2} \, d\phi \, d\phi'$$

$$\geq \frac{1}{4}c_\gamma \int_0^{4\pi}\int_0^{4\pi} \frac{[x(\phi) - x(\phi')]^2}{[\phi - \phi']^2} \, d\phi \, d\phi'$$

$$\geq \frac{1}{4}c_\gamma \int_0^{4\pi}\int_0^{4\pi} \frac{|x(\phi) - x(\phi')|^2}{|\phi - \phi'|^2} \, d\phi \, d\phi' - \frac{1}{4}c_\gamma \int_0^{4\pi}\int_0^{4\pi} \left(\frac{4\pi}{\phi_x}\right)^2 \, d\phi \, d\phi',$$

where

(2.7) $\phi_x = \max\{\overline{\phi} > 0 | \, |x(\phi + \overline{\phi}) - x(\phi)| < \pi, \; \forall \phi\} > 0$

is a constant depending on (the modulus of continuity of) x.

Therefore, for monotone x we have $E(x) < \infty$ iff $x \in \{\,\text{id}\,\} + T$.

$\square$

Remark 2.4:   The proof of Lemma 2.3 implies that the mapping X given by (2.1) between M with the topology induced from T and $C_0(\Gamma)$ is bounded. Moreover, for any uniformly convergent sequence $\{x_m\}$ of parametrizations $x_m \in M$ the constant ϕ_{x_m} given in (2.7) is uniformly bounded away from 0 . Hence if $E(x_m) \leq c < \infty$ uniformly, also $|x_m|_{1/2}$ will be uniformly bounded.

From now on we shall always endow the set M with the topology induced by the inclusion

$$M \subset \{\,\text{id}\,\} + T.$$

Similarly, $C_0(\Gamma)$ will be endowed with the $H^{1,2} \cap L^\infty$ −topology.

Lemma 2.5:   The map $X : M \to C_0(\Gamma)$ given by (2.1) extends to a differentiable map of the affine Banach space $\{\,\text{id}\,\} + T$ into $H^{1,2} \cap L^\infty(B; \mathbb{R}^n)$ of class C^{r-1}, if $\gamma \in C^r$, $r \geq 2$.

Proof: Note that for any $X \in H^{1,2} \cap L^\infty(B; \mathbb{R}^n)$ there is a unique harmonic surface $X_o \in H^{1,2} \cap L^\infty(B, \mathbb{R}^n)$ which agrees with X on ∂B in the sense that $X_o \in X + H_o^{1,2}(B; \mathbb{R}^n)$. Indeed, X_o is characterized by the variational principle

$$D(X_o) = \inf\{D(X') | X' \in X + H_o^{1,2}(B; \mathbb{R}^n)\}.$$

Existence of X_o follows easily from Theorem I.3.2; necessarily X_o is harmonic; boundedness and uniqueness are a consequence of the maximum principle.

Now "define" the trace space $H^{1/2,2}(\partial B; \mathbb{R}^n) \hat{=} H^{1,2}(B; \mathbb{R}^n)/H_o^{1,2}(B; \mathbb{R}^n)$ to be the set of equivalence classes $X|_{\partial B} \hat{=} X + H_o^{1,2}(B; \mathbb{R}^n)$ endowed with the quotient topology. In particular, let

$$\frac{1}{2}\,|X|_{\partial B}|_{1/2}^2 = \inf\{D(X') | X' \in X + H_o^{1,2}(B; \mathbb{R}^n)\} = D(X_o).$$

be the semi-norm on $H^{1/2,2}(\partial B; \mathbb{R}^n)$. In fact $H^{1/2,2}(\partial B; \mathbb{R}^n)$ is a Hilbert space with respect to the scalar product induced by $L^2(\partial B; \mathbb{R}^n)$ and the bilinear map

$$(X|_{\partial B}, Y|_{\partial B})_{1/2} = (X_o, Y_o)_1 = \int_B \nabla X_o \nabla Y_o \, dw,$$

where X_o and Y_o are the unique harmonic extensions of $X_{\partial B}$, $Y|_{\partial B}$ resp. By construction, the harmonic extension h is a linear isomorphism from $H^{1/2,2} \cap$

$L^\infty(\partial B;\ I\!R^n)$ into $H^{1,2} \cap L^\infty(B;\ I\!R^n)$. Moreover, by (2.4) an intrinsic definition of the semi-norm $|\cdot|_{1/2}$ in $H^{1/2,2}$ can be given in terms of Douglas' integral

$$D(X) = \frac{1}{16\pi} \int\limits_0^{2\pi} \int\limits_0^{2\pi} \frac{|X(e^{i\phi}) - X(e^{i\phi'})|^2}{\sin^2\left(\dfrac{\phi - \phi'}{2}\right)} d\phi\, d\phi',$$

which is equivalent to the usual definition

$$(2.8) \qquad |X|_{\partial B}|^2_{1/2} = \int\limits_{\partial B} \int\limits_{\partial B} \frac{|X(w) - X(w')|^2}{|w - w'|^2} dw\, dw',$$

cf. Adams [1, Theorems 7.48, 7.53].

It remains to show that the map $x \mapsto \gamma \circ x$ given by $(2.1'')$ extends to a differentiable map of class C^{r-1} of $\{\text{id}\} + T$ into $H^{1/2,2} \cap L^\infty(\partial B; I\!R^n)$ if γ is a diffeomorphism of class C^r, $r \geq 2$. But this directly follows from the chain rule, the pointwise representation

$$\frac{d^s}{dx^s}(\gamma \circ x)(\xi_1, \ldots, \xi_s) = \left(\left(\frac{d^s\gamma}{d\phi^s}\right) \circ x\right) \cdot \xi_1 \cdot \ldots \cdot \xi_s,$$

and the following elementary

Lemma 2.6:   i) For any $\sigma \in C^1(I\!R^m;\ I\!R^n)$, $\rho \in H^{1/2,2} \cap L^\infty(\partial B; I\!R^m)$ the composition $\sigma \circ \rho \in H^{1/2,2} \cap L^\infty(\partial B : I\!R^n)$ and

$$|\sigma \circ \rho|_{1/2} \leq \|(\nabla\sigma) \circ \rho\|_\infty \cdot |\rho|_{1/2}.$$

ii) For any $\xi, \eta \in H^{1/2,2} \cap L^\infty(\partial B)$ the product $\xi \cdot \eta \in H^{1/2,2} \cap L^\infty(\partial B)$ and

$$|\xi \cdot \eta|_{1/2} \leq \|\xi\|_\infty\, |\eta|_{1/2} + \|\eta\|_\infty\, |\xi|_{1/2}.$$

iii) For any $\xi \in H^{1/2,2}(\partial B)$, $\eta \in C^1(\partial B)$ the product $\xi \cdot \eta \in H^{1/2,2}(\partial B)$ and

$$|\xi \cdot \eta|_{1/2} \leq \|\xi\|_0\, \|\frac{d}{d\phi}\eta\|_\infty + \|\eta\|_\infty\, |\xi|_{1/2}.$$

The **proof** of Lemma 2.6 is immediate from (2.8). Hence also the proof of Lemma 2.5 is complete.

$$\square$$

In particular, from Lemma 2.5 we obtain

Lemma 2.7:   The functional $E : M \to I\!R$ extends to a functional of class C^{r-1} on $\{\text{id}\} + T$, if $\gamma \in C^r$, $r \geq 2$.

Proof: $E(x) = D(X(x))$, and D is a quadratic (analytic) functional.

□

In order to apply the abstract tools developed in the preceding section we will need a compactness property like Proposition 2.1 . However, for reasons that will become clear in Section II.4, we prefer to introduce a different normalization with respect to the conformal group action than the three-point-condition introduced earlier.

Let G be our representation (I.4.1) of the conformal group of the disc with tangent space

$$T_{\mathrm{id}}G = span\{iw,\ 1 - w^2,\ i(1 + w^2)\}.$$

In polar coordinates $w = re^{i\phi}$ an element $g \in G$ is represented by a map $\hat{g}$

$$g(w) = e^{i\vartheta(\phi)},\ \forall\ w = e^{i\phi} \in \partial B,$$

and similarly $T_{\mathrm{id}}G$ is represented

$$T_{\mathrm{id}}G = span\{1, \sin\phi, \cos\phi\}.$$

In the following, for ease of notation we will no longer distinguish between a conformal map $g \in G$ and the associated mapping $\hat{g}$.

Define

$$T^\dagger = \{\xi \in T \mid \int_0^{2\pi} \xi \cdot \eta\, d\phi = 0,\ \forall \eta \in T_{id}G\}$$

and let

$$M^\dagger = M \cap \left(\{id\} + T^\dagger\right),$$

endowed with the topology inherited from T .

Analogous to Proposition 2.1 we have:

Lemma 2.8 :   For any $\alpha \in I\!R$ the set

$$\{x \in M^\dagger \mid E(x) \leq \alpha\}$$

is compact with respect to the C^0-topology.

Proof :   Let $\{x_m\}$ be a sequence in $M^\dagger$ such that $E(x_m) \leq \alpha$ for some $\alpha \in I\!R$ and uniformly in m. By monotonicity and periodicity we have

$$x_m(0) \leq \frac{1}{2\pi} \int_0^{2\pi} x_m\, d\phi = \pi \leq x_m(2\pi) = x_m(0) + 2\pi .$$

Hence $|x_m(0)| \leq \pi$, uniformly, and Proposition 2.2 implies that $x_m \to x$ in $L^1_{loc}(\mathbb{R})$ and either

$$x(\phi) \equiv const \ (mod \ 2\pi)$$

or $x \in M$ and

$$x_m \to x \text{ uniformly as } m \to \infty .$$

Moreover, in the second event, by lower semi-continuity of E with respect to uniform convergence in M, $E(x) \leq \alpha$, and the proof is complete in this case.

To rule out the first possibility we introduce $\xi_m = x_m - id$, $\xi = x - id$ and use the normalization conditions

$$\int\limits_0^{2\pi} \xi_m \cdot \eta \ d\phi = o, \ \forall \eta \in T_{id}G .$$

Since $x_m \to x$ in L^1 , also $\xi_m \to \xi$ in L^1 , and also ξ satisfies these constraints which we may write down explicitly as follows:

$$\int\limits_0^{2\pi} \xi \ d\phi = \int\limits_0^{2\pi} \xi \ \sin \phi \ d\phi = \int\limits_0^{2\pi} \xi \ \cos \phi \ d\phi = 0 .$$

But if $x \equiv const \ (mod \ 2\pi)$, and is 2π-periodic and monotone, ξ is of the form

$$\xi(\phi - \phi_0) = c - \phi, \ 0 \leq \phi \leq 2\pi .$$

The first condition now implies $c = \pi$, while the second and third condition give

$$0 = \int\limits_0^{2\pi} \xi(\phi) \left(\sin \phi \ \cos \phi_0 + \cos \phi \ \sin \phi_0 \right) d\phi$$

$$= \int\limits_0^{2\pi} \xi(\phi) \ \sin(\phi + \phi_0) \ d\phi = \int\limits_0^{2\pi} \xi(\phi - \phi_0) \ \sin(\phi) \ d\phi = -\int\limits_0^{2\pi} \phi \ \sin \phi \ d\phi > 0 .$$

The contradiction shows that the case $x \equiv const \ (mod \ 2\pi)$ does not occur. The proof is complete.

□

For $x \in M^\dagger$ let

$$g(x) = \sup_{\substack{y \in M^\dagger \\ |x - y| < 1}} \langle dE(x), \ x - y \rangle$$

as in Definition 1.1.

Our approach is meaningful since we can identify critical points of E on $M^\dagger$ with minimal surfaces spanning Γ by means of the following

Proposition 2.9:   A point $x \in M^\dagger$ is critical for E, i.e. $g(x) = 0$, iff the surface $X = X(x)$ is a minimal surface spanning Γ.

Proof: Compute

$$\langle dE(x), \xi \rangle = \frac{d}{d\epsilon} D(X(x + \epsilon\xi))|_{\epsilon=0}$$

(2.9)
$$= \int_B \nabla X(x) \nabla(\frac{d}{d\epsilon} X(x + \epsilon\xi)|_{\epsilon=0}) dw$$

$$= \int_{\partial B} \partial_n X \frac{d}{d\epsilon} (\gamma \circ (x + \epsilon\xi))|_{\epsilon=0} do$$

$$= \int_{\partial B} \partial_n X \cdot \frac{d}{d\phi}\gamma(x) \cdot \xi \, do, \ \forall \, x \in M, \, \xi \in T.$$

If X solves (1.1) - (1.3), by Theorem 5.1 and since $\gamma \in C^r$, $r \geq 2$, $X \in C^1(\overline{B}; \mathbb{R}^n)$. Thus $\partial_n X \cdot \frac{d}{d\phi}\gamma(x) \equiv 0$ on ∂B by (1.2) and $dE(x) = 0$.

To prove the converse implication for simplicity let us for the moment assume the following regularity result:

Proposition 2.10:   If $X \in M^\dagger$ solves $g(x) = 0$, and $\gamma \in C^r$, $r \geq 2$, then $X = X(x) \in H^{2,2}(B; \mathbb{R}^n)$.

Proposition 2.10 along with other regularity results will be established in Section 5 of this chapter.

Proof of Proposition 2.9 (continued):   If $X \in H^{2,2}(B; \mathbb{R}^n)$ by the trace theorem, cp. Adams [1; Thm. 7.53, 7.57] , we have
(2.10')
$$\partial_n X|_{\partial B} \, , \, \frac{d}{d\phi} X|_{\partial B} = ((\frac{d}{d\phi}\gamma) \circ x) \cdot \frac{d}{d\phi} x \in H^{1/2 ,2}(\partial B; \mathbb{R}^n) \hookrightarrow L^p(\partial B; \mathbb{R}^n), \ \forall p < \infty.$$

But γ is a diffeomorphism. Hence also

(2.10")
$$\frac{d}{d\phi} x \in L^p(\partial B; \mathbb{R}^n), \ \forall p < \infty.$$

For $\xi \in C^1(\mathbb{R}/2\pi)$ such that $\|\frac{d}{d\varphi}\xi\|_\infty < 1$ the map

$$\phi \to \phi + \xi(\phi)$$

induces a diffeomorphism of $\partial B \cong \mathbb{R}/2\pi$.

Fix $\xi \in C^1(\mathbb{R}/2\pi)$, and for sufficiently small ϵ and $\eta \in T_{id}G \subset C^1(\mathbb{R}/2\pi)$ consider the parametrizations

$$x(\epsilon, \eta) = x \circ (id + \epsilon\xi + \eta) .$$

The map $F : \mathbb{R} \times T_{id}G \to \mathbb{R}^3$ given by

$$(\epsilon, \eta) \to \left(\int_0^{2\pi} x(\epsilon, \eta)\, d\phi, \ \int_0^{2\pi} x(\epsilon, \eta)\, \sin\phi\, d\phi, \ \int_0^{2\pi} x(\epsilon, \eta)\, \cos\phi\, d\phi \right)$$

being continuously differentiable with $\frac{\partial F}{\partial \eta}(0,0) : T_{id}G \to \mathbb{R}^3$ surjective, by the implicit function theorem there exists a differentiable mapping $\epsilon \mapsto \eta(\epsilon)$ in a neighborhood of $\epsilon = 0$, $\eta(0) = 0$ such that

$$x_\epsilon = x \circ (id + \epsilon\, \xi + \eta(\epsilon)) \in M^\dagger$$

for all ϵ. Now, $dE(x)$ by (2.10) continuously extends to a functional on $L^2(\partial B)$. Hence, differentiating by the chain rule and using the conformal invariance of E we obtain

$$\frac{d}{d\epsilon} E(x_\epsilon)\big|_{\epsilon = 0^+} = \int_{\partial B} \partial_n X \cdot \frac{d}{d\phi}\gamma(x) \cdot \left(\frac{d}{d\phi} x \cdot \xi \right) do$$

$$= \langle dE(x), \frac{d}{d\epsilon} x_\epsilon \rangle = - \lim_{\epsilon \to 0^+} \frac{1}{\epsilon} \langle dE(x), x - x_\epsilon \rangle \geq 0.$$

Note that we have used the fact that $g(x) = 0$. Hence, reversing the sign of ϵ, we obtain that

$$(2.11) \qquad \int_{\partial B} \partial_n X \cdot \frac{d}{d\phi}\gamma(x) \cdot \frac{d}{d\phi} x \cdot \xi \, do = \int_{\partial B} \partial_n X \cdot \frac{d}{d\phi} X \cdot \xi \, do = 0$$

for all $\xi \in C^1(\mathbb{R}/2\pi)$, and by density of such ξ in $L^2(\partial B)$:

$$\partial_n X \cdot \frac{d}{d\phi} X = 0 \quad \text{on } \partial B.$$

Now recall that by Lemma I.2.3 the function

$$\hat{\Phi}(w) = w^2 (\partial X)^2$$

$$= \left[(u + iv) \left(\frac{\partial}{\partial u} X - i\frac{\partial}{\partial v} X \right) \right]^2$$

$$= \left[r\frac{\partial}{\partial r} X - i\frac{\partial}{\partial \phi} X \right]^2$$

$$= r^2 \left| \frac{\partial}{\partial r} X \right|^2 - \left| \frac{\partial}{\partial \phi} X \right|^2 - 2ir\frac{\partial}{\partial r} X \cdot \frac{\partial}{\partial \phi} X$$

is a holomorphic function of $w = u + iv = re^{i\phi}$ on B. Its imaginary part vanishing on ∂B, necessarily $\hat{\Phi}$ is real on B and hence constant by the Cauchy - Riemann equations. Inspection at $r = 0$ now yields that

$$\hat{\Phi} \equiv \hat{\Phi}(0) = 0,$$

whence also
$$\Phi(w) = (\partial X)^2 = |X_u|^2 - |X_v|^2 - 2iX_u \cdot X_v \equiv 0$$
and X is conformal.

□

Let us verify the Palais-Smale condition (P.S.) for E on $M^\dagger$.

Lemma 2.11:   Any sequence $\{x_m\}$ in $M^\dagger$ such that $E(x_m) \leq c$ uniformly while $g(x_m) \to 0$ as $m \to \infty$ is relatively compact.

Proof: By Lemma 2.8 and Remark 2.4 a subsequence $x_m \to x \in M$ uniformly, while $\{x_m\}$ is equi-bounded in $H^{1/2,2}$. Moreover, letting $X_m = X(x_m)$, $X = X(x)$, we may assume that also $X_m \xrightarrow{w} X$ weakly in $H^{1,2}(B; \mathbb{R}^n)$.

On ∂B , expand
$$X_m - X = \gamma \circ x_m - \gamma \circ x$$
$$= \frac{d}{d\phi}\gamma(x_m)(x_m - x) - \int\limits_x^{x_m} \int\limits_{x'}^{x_m} \frac{d^2}{d\phi^2}\gamma(x'')dx''dx'$$
$$=: \frac{d}{d\phi}\gamma(x_m)(x_m - x) + I_m.$$

Let $\sigma \in C^1(\mathbb{R}^2; \mathbb{R}^n)$ be given by
$$\sigma(y, z) = \int\limits_y^z \int\limits_{x'}^z \frac{d^2}{d\phi^2}\gamma(x'')dx''dx'$$

and let $\rho_m \in H^{\frac{1}{2},2}(\mathbb{R}/2\pi; \mathbb{R}^2)$ be the map
$$\rho_m(\phi) = (x(\phi), x_m(\phi)).$$

By Lemma 2.6.i) the composed map $\sigma \circ \rho_m \in H^{1/2,2} \cap L^\infty$ and
$$|I_m|_{1/2} = |\sigma \circ \rho_m|_{1/2} \leq ||(\nabla \sigma) \circ \rho_m||_\infty \cdot |\rho_m|_{1/2}$$
$$\leq C||x_m - x||_\infty(|x_m|_{1/2} + |x|_{1/2}).$$

Hence with error terms $o(1) \to 0$ $(m \to \infty)$
$$\int\limits_B |\nabla(X_m - X)|^2 dw = \int\limits_{\partial B} \partial_n X_m(X_m - X)do + o(1)$$
$$= \int\limits_{\partial B} \partial_n X_m \cdot \frac{d}{d\phi}\gamma(x_m)(x_m - x)do + o(1)$$
$$= \langle dE(x_m), x_m - x \rangle + o(1) \leq g(x_m)|x_m - x| + o(1)$$
$$\to 0 \ (m \to \infty).$$

Thus $X_m \to X$ in $H^{1,2}(B; \mathbb{R}^n)$ and Lemma 2.5 implies that $x_m \to x$ strongly in $M^\dagger$ as $m \to \infty$.

$\square$

Theorems 1.12, 1.13 now immediately imply the main result of this section, the "mountain-pass-lemma for minimal surfaces" due to Morse-Tompkins [1] and Shiffman [1] .

Actually, the result below is a slight improvement of the Morse-Shiffman-Tompkins results in as much as our Theorem 1.13 also enables us to handle the case of relative minima which are not necessarily strict.

Theorem 2.12:   Suppose a Jordan curve Γ of class C^2 bounds two distinct relative minima $X_1, X_2 \in \mathcal{C}(\Gamma)$. Then either $D(X_1) = D(X_2) = \beta_0$ and X_1, X_2 can be connected in any neighborhood of the set of minimal surfaces X spanning Γ with $D(X) = \beta_0$, or there exists a minimal surface X_3 spanning Γ distinct from X_1, X_2 which is not a relative minimum for D on $\mathcal{C}(\Gamma)$, i.e. which is an unstable solution of (1.1) - (1.3).

Remark 2.13:   In particular, if $\Gamma \in C^2$ bounds two distinct relative minima $X_1, X_2 \in \mathcal{C}^*(\Gamma)$ at least one of which is strict, then Γ also bounds an unstable minimal surface.

Proof of Theorem 2.12 :   By conformal invariance we may assume that $X_i = X(x_i)$ with $x_i \in M^\dagger$, $i = 1, 2$. Then x_1, x_2 are relative minima of E , and by Theorem 1.13 and Lemma 2.11 either $E(x_1) = E(x_2) = \beta_0$ and x_1, x_2 are connected in any neighborhood of K_{β_0} in $M^\dagger$ (hence in M), or there exists a critical point x_3 which is not a relative minimum of E in $M^\dagger$ (hence not in M , either). Composing with $X : M \to \mathcal{C}(\Gamma)$ we obtain the Theorem.

$\square$

3. Morse theory on convex sets.   More detailed information about the number and types (relative minimum, saddle-point) of critical points of functionals can be obtained from Morse theory. In the 20's Marston Morse [1] developed a method for relating properties of the set of critical points of a functional with global topological properties of the space over which the functional is defined. Morse' ideas were recast by Milnor [1] in a more applicable form. A major extension of Milnor's approach to Morse theory was then made by Palais[1],[2], Palais-Smale [1] and Smale [1], in the 60's.

Today a theory is emerging inspired by the deep work of Charles Conley [1] which seeks to combine the generality of Morse' original theory with the clarity of Milnor's approach. However, inspite of encouraging results by Rybakowski [1], Conley - Zehnder [1], and Rybakowski - Zehnder [1], Conley's version of Morse theory has not yet been conveyed to infinite-dimensional spaces.

With applications to minimal surfaces in mind we proceed to develop a Morse theory for functionals defined on convex sets.

Our presentation-based on Struwe [1] -to a large extent uses ideas and notions introduced by Milnor [1] and Palais [1]. However, in order to avoid coordinate transformations (which might destroy the convexity of M) normal form representations (the "Morse Lemma") are replaced by the use of pseudo - gradient vector fields, cf. Definition 1.7. Besides reducing regularity requirements on E this approch also seems more natural and simpler conceptually.

Throughout this section we make the following

Assumption:

Let M be a closed convex subset of an (affine) Banach space T with norm $|\cdot|_T$ which is densely and continuously embedded in a Hilbert space H with inner product $(\cdot,\cdot)$ and induced norm $|\cdot|_H$

$$(3.1) \qquad \underset{\substack{\text{closed} \\ \text{convex}}}{M \subset} \overset{\text{dense}}{T \hookrightarrow H.}$$

We moreover assume that M is also closed in H and that the topologies induced on M by H and T coincide. Also suppose that $E : M \to \mathbb{R}$ extends to a C^2-functional on T and that at any critical point $x_o \in M$-in the sense of Definition 1.3- the form $d^2E(x_o)$ extends continuously to $H \times H$ with expansions for E and dE :

$$(3.2') \qquad E(x) = E(x_o) + \frac{1}{2} d^2E(x_o)(x - x_o, \ x - x_o) + o(|x - x_o|_H^2),$$

$$(3.2'') \qquad \langle dE(x), \ x - y \rangle = d^2E(x_o)(x - x_o, \ x - y) + o(|x - x_o|_H^2)$$

for all $x, y \in M$ such that $|y - x|_H \leq |x - x_o|_H$.

Remark 3.1:   i)   In case $M = T = H$ our assumptions simply mean that $E \in C^2(H)$.

ii)   Note that by (3.2) implicitly we assume that at any critical point x_o of E in M there holds $\langle dE(x_o), x_o - x \rangle = 0$, $\forall x \in M$.

iii)   Actually, $E \in C^1(T)$ and the existence of $d^2 E(x_o)$ at any critical point of E on M, together with the expansions (3.2) would suffice.

iv)   We do *not* assume that (3.2) holds for $x \notin M$.

By the Riesz representation theorem and (3.2) at any critical point $x_o \in M$ there exists a bounded self-adjoint linear map $a(x_o) : H \to H$ such that

$$(a(x_o)\xi, \eta) = d^2 E(x_o)(\xi, \eta), \ \forall \, \xi, \eta \in H.$$

Let H_0 be the kernel of $a(x_o)$, and let H_-, H_+ be maximal $a(x_o)$-invariant subspaces of H such that

$$(a(x_o)\xi, \xi) < 0 \quad , \forall \, \xi \in H_-, \ \xi \neq 0,$$
$$(a(x_o)\xi, \xi) > 0 \quad , \forall \, \xi \in H_+, \ \xi \neq 0.$$

This defines a decomposition $H = H_+ \oplus H_o \oplus H_-$ which is orthogonal with respect to $(\cdot, \cdot)$ and $d^2 E(x_o)$, our *standard decomposition* of H at a critical point $x_o \in M$. (In case $a(x_o)$ has a pure point-spectrum, H_+, H_- are spanned by eigenvectors of $a(x_o)$ corresponding to positive, resp . negative eigenvalues of $a(x_o)$.)

Definition 3.2: $x_o \in M$ is a *non-degenerate* critical point of E if $a(x_0)$ is a topological isomorphism of H. In this case the *Morse index* of x_o is given by

$$\text{index} \, (x_o) = \quad \dim H_-,$$

where $H = H_+ \oplus H_o \oplus H_-$ denotes the standard decomposition of H at x_o.

We now add as a further *assumption* :

At any critical point x_o there is a neighborhood U_- of 0 in H_- such that

(3.3) $$\{x_0\} + U_- \subset M.$$

Before we state the main theorem of this section we need two more definitions.

Definiton 3.3:   Let S, T be topological spaces. T arises from S by attachment ψ of a handle h of type r iff

i) $T = S \cup h$

ii) $\psi : \overline{B}_1(0; I\!\!R^r) \rightarrow h$ is a homeomorphism, and also the restrictions

$$\psi|_{\partial B_1(0; I\!\!R^r)} : \partial B_1(0; I\!\!R^r) \rightarrow S \cap \partial h.$$

$$\psi|_{B_1(0; I\!\!R^r)} : B_1(0; I\!\!R^r) \rightarrow T \backslash S$$

are homeomorphisms.

Definition 3.4:   Let S, T be topological spaces. S and T are homotopically equivalent iff there are continuous mappings $f : S \rightarrow T$, $g : T \rightarrow S$ such that the composition

$f \circ g : T \rightarrow T$ is homotopic to the identity on T

and

$g \circ f : S \rightarrow S$ is homotopic to the identity on S.

The maps f and g in this case are homotopy equivalences.

Remark 3.5:   i) Definition 3.4 defines an equivalence relation on the category of topological spaces and maps.

ii) Note that certain topological properties of spaces are invariant under homotopy equivalence: If $+ : S \rightarrow G(S)$ is a map from the category of topological spaces and maps to the category of groups and group homomorphisms which is functorial in the sense that

-     Any $f : S \rightarrow T$ induces a homomorphism $f^+ : G(S) \rightarrow G(T)$ depending only on the homotopy class of f .

-     If $f : S \rightarrow T$, $g : T \rightarrow U$, then $(g \circ f)^+ = g^+ \circ f^+$.

-     The identity $1_S : S \rightarrow S$ induces an automorphism $1_S^+ : G(S) \rightarrow G(S)$.

then the structure group $G(S)$ up to automorphisms will be independent of the particular representant in the homotopy equivalence class of S. In particular, the ranks of the homology groups of S will be invariant under homotopy equivalence,cf. Djugundji [1].

For a functional E on M satisfying the above hypotheses (3.1)-(3.3) and possessing only non-degenerate critical points, let

$$C_m = |\{x \in M | g(x) = 0, \quad \text{index}(x) = m\}|$$

denote the number of critical points of E on M with index m.

Then we obtain the following

Theorem 3.6:   Suppose E is a functional on M satisfying assumptions (3.1) - (3.3) and possessing only non-degenerate critical points. Let the Palais-Smale condition be satisfied for E on M.

i) Then for any constant $c \in I\!R$ there are only finitely many critical points x of E with $|E(x)| \le c$.

ii) Moreover, if for $\alpha < \beta < \gamma$ the number β is the only critical value of E in $[\alpha, \gamma]$, and if $x_1, ..., x_k$ are the only critical points of E with $E(x_j) = \beta$ having Morse indeces $r_1, ..., r_k$, then M_γ is homotopically equivalent to M_α with k handles of types $r_1, ..., r_k$ disjointly attached.

iii) If the energies of critical points of E are uniformly bounded, then E possesses only finitely many critical points and the Morse relations hold:

(3.4 ')
$$C_0 \ge 1,$$
$$\sum_{m=0}^{l} (-1)^{l-m} C_m \ge (-1)^l,$$
$$\sum_{m=0}^{l} (-1)^m C_m = 1.$$

Remark 3.7:   The Morse relations (3.4') may be summarized in the single identity

(3.4 ")
$$\sum_{m=0}^{\infty} C_m t^m = 1 + (1+t)Q(t)$$

where Q is a polynominal with non-negative integer coefficients, and the polynominal $\sum_{m=0}^{\infty} R_m t^m \equiv 1$ is the Poincaré polynominal of M, cf. Rybakowski - Zehnder [1, p.124], the numbers

$$R_m = \operatorname{rank}(H_m(M)) = \begin{cases} 1, & m = 0 \\ 0, & m > 0 \end{cases}.$$

denoting the Betti numbers of the (convex hence) contractible space M.

Proof: By (3.2) and our non-degeneracy assumption critical points are isolated. By (P.S.) the set of critical points having uniformly bounded energy is compact, hence finite if it consists of isolated points. This proves i) and the first part of iii).

Postponing the proof of ii) for a moment let us derive the Morse inequalities (3.4). Let $\beta_1 < ... < \beta_j$ be the critical values of E and choose regular values α_i, γ_i such that

$$\alpha_1 < \beta_1 < \gamma_1 = \alpha_2 < \beta_2 < ... \beta_j < \gamma_j.$$

For each pair of regular values α, γ let

$$R_m^{\alpha,\gamma} = \operatorname{rank}(H_m(M_\gamma, M_\alpha))$$

be the Betti numbers of the pair (M_γ, M_α), and let

$$C_m^{\alpha,\gamma} = |\{x \in M_\gamma \backslash M_\alpha | g(x) = 0, \; \text{Index}(x) = m\}|.$$

By ii) for each pair α_i, γ_i M_{γ_i} is homotopically equivalent to M_{α_i} with k_i handles of types $r_1^i, ... r_{k_i}^i$ disjointly attached, where $r_1^i, ..., r_{k_i}^i$ are the Morse indices of the critical points of E at energy β_i. By Remark 3.5. ii) therefore the Betti numbers of $(M_{\gamma_i}, M_{\alpha_i})$ are the same as those of a disjoint union of k_i pointed spheres (S^d, p) of dimensions $d = r_1^i, ..., r_{k_i}^i$.

Since

$$\text{rank}\left(H_m\left(S^d, p\right)\right) = \begin{cases} 1 & , m = d \\ 0 & , \text{else,} \end{cases}$$

we obtain the relations

$$R_m^{\alpha_i, \gamma_i} = C_m^{\alpha_i, \gamma_i}, \quad 1 \le i \le j.$$

Adding, cycles may cancel while critical points cannot and we obtain (cf. Palais [1, p. 336 ff.]) the system of inequalities for all regular $\alpha < \gamma$:

$$
\begin{aligned}
&R_m^{\alpha,\gamma} \le C_m^{\alpha,\gamma}, \; \forall\, m \in I\!N_o \\
&\sum_{m=0}^{l} (-1)^{l-m} R_m^{\alpha,\gamma} \le \sum_{m=0}^{l} (-1)^{l-m} C_m^{\alpha,\gamma}, \; \forall\, l \in I\!N_o \\
&\sum_{m=0}^{\infty} (-1)^{m} R_m^{\alpha,\gamma} = \sum_{m=0}^{\infty} (-1)^{m} C_m^{\alpha,\gamma}.
\end{aligned}
$$
(3.5)

Equality in the last line corresponds to the well-known additivity of the Euler characteristic. Letting $\alpha \to -\infty$, $\gamma \to \infty$ the right hand sides of (3.5) stabilize for large α, γ while the quantities on the left for large α, γ are bounded from below by the corresponding expressions involving the Betti numbers R_m of M. This completes the proof of (3.4).

It remains to establish ii).

Preliminaries: Let $\alpha < \gamma$ be regular values of E and for simplicity assume that $x_o \in M$ is the only critical point of E in M having $E(x_o) = \beta \in [\alpha, \gamma]$. Let r_o be the index of x_o, $H = H_+ \oplus H_-$ the standard decomposition of H at x_o. For any $\xi \in H$ denote $\xi = \xi_+ + \xi_- \in H_+ \oplus H_-$ its components.

Choose $0 < \rho < 1$ such that

(3.6) $$\{x_o\} + B_{2\rho}(0; H_-) \subset M$$

which is possible by assumption (3.3).

By non-degeneracy of x_0 there is a constant $\lambda > 0$ such that

(3.7 ') $$|d^2 E(x_o)(\xi, \xi)| \ge \lambda |\xi|_H^2, \; \forall \xi \in H.$$

By (3.2) we may suppose that ρ is chosen such that

(3.7 ") $\qquad |\langle dE(x), \, x - y \rangle - d^2 E(x_o)(x - x_o, x - y)| \leq \frac{\lambda}{2}|x - x_o|_H^2$

for all $x \in B_{2\rho}(x_o, H) \cap M$, all $y \in M$, provided $|x - y|_H \leq |x - x_o|_H$. In particular, the vector field e_o given by

$$e_o(x) = (x - x_o)_- - (x - x_o)_+$$

is a pseudo-gradient vector field for E on $U := B_\rho(x_o, H) \cap M$ in the following sense: By (3.6) for any $x \in U$

$$e_o(x) + x = 2(x - x_o)_- - (x - x_o) + x = x_o + 2(x - x_o)_- \in M.$$

Moreover,

$$|e_o(x)|_H = |x - x_o|_H \leq \rho < 1 \, ,$$

while by (3.7)

$$g(x) \leq c \cdot \sup_{\substack{y \in M \\ |x-y|_H < 1}} \langle dE(x), x - y \rangle \leq c \cdot \left(\|d^2 E(x_o)\| + \frac{\lambda}{2} \right) |x - x_o|_H,$$

whence

$$\langle dE(x), e_o(x) \rangle \leq -\frac{\lambda}{2}|x - x_o|_H^2 \leq -c^{-1} g(x)^2,$$

with a uniform constant $c \in I\!R$.

Finally, e_o is Lipschitz continuous in the H−norm.

Now let ψ_o, $0 \leq \psi_o \leq 1$, be a Lipschitz continuous function with support in U and $\equiv 1$ in a neighborhood $V \subset M_\gamma \backslash M_\alpha$ of x_o. Let $\tilde{e} : \tilde{M} \to T \subset H$ be a pseudo-gradient vector field as constructed in Lemma 1.8 and Lipschitz continuous in H. Remark that by assumption (3.1) the function g – being continuous in T, cp. Remark 1.2.ii) – is also continuous in H. Hence the construction of Lemma 1.8 conveys. Then the vector field

$$e(x) = \psi_o(x) e_o(x) + (1 - \psi_o(x)) \tilde{e}(x)$$

defined in a neighborhood of $M_\gamma \backslash M_\alpha$ will have the following properties:

(3.8)
$$\begin{aligned} &e \text{ is Lipschitz continuous in } H; \\ &e(x) + x \in M, \quad \forall x \in M_\gamma \backslash M_\alpha, \\ &|e(x)|_H \leq c, \\ &\langle dE(x), \, e(x) \rangle \leq -\min\{c^{-1} g^2(x), \, 1\} \end{aligned}$$

for some uniform constant $c \in I\!R$.

As in the proof of Proposition 1.9 we may now construct a deformation Φ of a neighborhood of $M_\gamma \backslash M_\alpha$ by letting Φ be the unique solution to the initial value problem

(3.9) $\qquad \dfrac{\partial}{\partial t}\Phi(t, x) = e(\Phi(t, x)),$

$$\Phi(0, x) = x.$$

Indeed, by Lipschitz continuity of e Φ exists locally. Moreover, by uniform boundedness $|e|_H \le c$ and since by assumption M is also closed as a subset of H the trajectories $t \mapsto \Phi(t, x)$ are (relatively) closed and Φ can be continued globally as long as $\Phi(t, x)$ remains in a neighborhood of $M_\gamma \backslash M_\alpha$ where e is defined.

Since E is non-increasing along trajectories of Φ by construction, through any $x \in M_\gamma \backslash M_\alpha$ the solution $\Phi(t, x)$ of (3.9) will be defined for

(3.10) $0 < t < t(x) = \min\{t \ge 0 | E(\Phi(t, x)) \le \alpha\}.$

Moreover, since α is regular Φ meets M_α transversally, and the time-map $x \mapsto t(x)$ is continuous.

In the following we replace Φ by the localized flow $\Phi^{\alpha, \gamma} : [0, \infty[\times M_\gamma \to M_\gamma$ given by

$$\Phi^{\alpha, \gamma}(t, x) \;=\; \begin{cases} \Phi(\min\{t, t(x)\}, x) & , x \in M_\gamma \backslash M_\alpha \\ x & , x \in M_\alpha. \end{cases}$$

Note that $\Phi^{\alpha, \gamma}$ is continuous. This understood from now on for ease of notation let us again simply write Φ instead of $\Phi^{\alpha, \gamma}$.

The construction of the desired homotopy equivalence of M_γ with M_α with a handle of type r_o attached is now completed in three stages.

Step 1:

Choose a neighborhood W of x_o in M of the form $W = (\{x_o\} + B_\rho(0; H_+) \times B_\sigma(0; H_-)) \cap M$ such that $\overline{W} \subset V \subset M_\gamma$. In particular W is convex and the orthogonal projection of W onto $\{x_o\} + H_-$ coincides with

$$W_- = W \cap (\{x_o\} + H_-).$$

By Lemma 1.10 there exists a constant $\delta > 0$ such that for $x \in M_\gamma \backslash M_\alpha$, $x \notin W$ there holds $g(x) \ge \delta$. (3.8) then implies that for any $x \in M_\gamma$ either $\Phi(t, x)) \in M_\alpha \cup W$ for some t or

$$E(\Phi(t, x)) \;=\; E(x) \;+\; \int\limits_0^t \langle dE(\Phi(t, x)), \, e(\Phi(t, x)) \rangle dt$$

$$\le \gamma - t \; \min\{c^{-1}\delta^2, 1\}.$$

Hence for $t_0 = (\gamma - \alpha) \cdot \max\{1, c\delta^{-2}\}$ the flow Φ induces a homotopy equivalence

$$\Phi(t_o, \cdot) \; : \; M_\gamma \to M_\alpha \cup \Phi([o, t_o], \, W) \hookrightarrow M_\gamma.$$

Step 2:

Choose a Lipschitz continuous (in H) function ψ_1, $0 \leq \psi_1 \leq 1$, with support in V and $\equiv 1$ on W. For $x \in M$ let

$$x' = \psi_1(x)(x_o + (x - x_o)_-) + (1 - \psi_1(x))x.$$

Note that $x' \in M$ by (3.3). Now define a homotopy

$$\Psi : [0, t_0] \times M \to M \quad \text{by letting}$$
$$\Psi(t, x) = \frac{t}{t_o}x' + (1 - \frac{t}{t_o})x.$$

Observe that $\Psi(0, \cdot) = \mathrm{id}|_M$, $\Psi(t, \cdot)|_{M_\alpha} = \mathrm{id}|_{M_\alpha}$ for all $t \in [0, t_0]$ by choice of V.

Moreover, $\Psi(t_o, W) \subset W_-$. Also remark that since e (and hence) Φ is linear on V by construction and since W is convex we can achieve that

$$\Phi([0, t_o], \Psi(t, W)) \subset \Psi(t, \Phi([0, t_o], W)) \subset \Phi([0, t_o], W)$$

for all $t \in [0, t_o]$.

For this, since Φ is strictly E−decreasing by (3.8), we can choose W for fixed V so small that no trajectory $\{\Phi(t, x)|t \in [0, t_0]\}$ through $x \in W$ can re-enter W after leaving $\overline{W}$.
Hence the composition $\Phi \circ \Psi : [0, t_0] \times M_\gamma \to M_\gamma$ induces a homotopy equivalence

$$\Phi(t_o, \cdot) \circ \Psi(t_o, \cdot) \; : \; M_\alpha \cup \Phi([0, t_o], W) \to M_\alpha \cup \Phi([0, t_o], W_-).$$

Step 3:

To conclude the proof suppose $\{x_o\} + B_\sigma(0; H_-) \subset W_-$. For $\xi \in \overline{B}_1(0; I\!R^{r_o}) \subset I\!R^{r_o} \hat{=} H_-$ let

$$\psi(\xi) = \begin{cases} \Phi\left(t(x_o + \sigma \cdot \xi/|\xi|), \; x_o + \sigma \cdot \xi\right) & \xi \neq 0 \\ x_o & ,\xi = 0 \end{cases}$$

to obtain a homeomorphism ψ of $\overline{B}_1(0; I\!R^{r_o})$ onto $h := \Phi([0, t_o], W_-)$ satisfying the conditions of Definition 3.3. (The time-map $x \mapsto t(x)$ was defined in (3.10).)

$$\square$$

Remark 3.8 If $M = T = H$ then for any functional $E \in C^2(H)$ assumptions (3.1)-(3.3) are trivially satisfied. Since the above constructions are purely local Theorem 3.6 also conveys to C^2−functionals on Hilbert manifolds.

4. Morse inequalities for minimal surfaces.   We claim that the assumptions (3.1), (3.2) of the preceding section are satisfied if we let M be defined as in (2.2), $T = H^{1/2,2} \cap C^o(I\!R/2\pi)$ as in (2.5) , $H = H^{1/2,2}(I\!R/2\pi)$ and let $E(x)$ be given by (2.4).

Lemma 4.1: M is a closed convex subset of the affine Banach space $\{\,\text{id}\,\} + T$. T is a dense subspace of H, and the topologies induced by the inclusions $M \subset \{\,\text{id}\,\} + T$, $M \subset \{\,\text{id}\,\} + H$ coincide. Moreover, M is closed in $\{\,\text{id}\,\} + H$.

Proof: By Lemma 2.3 it remains to verify the last statements. Suppose by contradiction that $\{x_m\}$ is a sequence in M tending to x_o in H while $|x_m - x_o|_T \geq c > 0$ uniformly. Since the map X in (2.1) is bounded, cf. Remark 2.4, the energies $E(x_m)$ are uniformly bounded. By Proposition 2.2 $x_m \to x_0$ uniformly. (The case $x_0 = const.$ modulo 2π is excluded because $x_0 \in H^{1/2,2}$ and is monotone. In particular $x_0 \in M$ and M is closed in $\{\,\text{id}\,\} + H$; moreover, any relative T−neighborhood of a point $x_o \in M$ contains a relative H−neighborhood of x_o.

$\square$

Lemma 4.2:   Suppose $\gamma \in C^4$. Then E extends to a functional of class C^3 on $\{\,\text{id}\,\} + T$, and at any critical point $x_o \in M$ of E of M (in the sense of Definition 1.3) we have the expansion

$$E(x) = E(x_o) + \frac{1}{2}\, d^2 E(x_o)(x - x_o, x - x_o) + o(|x - x_o|_H^2),$$
$$\langle dE(x), x - y \rangle = d^2 E(x_o)(x - x_o, x - y) + o(|x - x_o|_H^2)$$

for all $x, y \in M$ such that $|y - x|_H \leq |x - x_o|_H$.

Proof: The first part follows from Lemma 2.7. Moreover, by Proposition 2.9 and Theorem I.5.1 a critical point $x_o \in M$ induces a minimal surface $X_o = X(x_o) \in C^3(\overline{B}, I\!R^n)$. Hence also $x_o \in C^3$.

Compute, using (2.9) for $\xi, \eta \in T$:

(4.1)
$$d^2 E(x_o)(\xi, \eta) = \int_{\partial B} \partial_n X \cdot \frac{d^2}{d\phi^2} \gamma(x_o) \cdot \xi \cdot \eta \, do$$
$$+ \int_B \nabla(dX(x_o) \cdot \xi) \cdot \nabla(dX(x_o) \cdot \eta) \, dw,$$

where

(4.2)
$$dX(x_o) \cdot \xi = h\left(\frac{d}{d\phi}\gamma(x_o) \cdot \xi\right).$$

By Lemma 2.6 and the above regularity properties of x_o, X_o the form $d^2E(x_o)$ continuously extends to $H \times H$.

To verify the expansion formula near x_o, first note that by (I.1.2), (2.9) at a critical point $x_o \in M$

$$dE(x_o) = 0 \in T^*.$$

Since $E \in C^3$ on $\{\,\mathrm{id}\,\} + T$ we hence obtain that

$$E(x) = E(x_o) + \frac{1}{2}\, d^2E(x_o)(x - x_o, x - x_o) + 0(|x - x_o|_T^3)$$

$$\langle dE(x), x - y \rangle = d^2E(x_o)(x - x_o, x - y) + 0(|x - x_o|_T^2)|x - y|_T.$$

Evidently

$$|x - x_o|_T^3 \le c(|x - x_o|_H^3 + ||x - x_o||_\infty^3),$$

$$|x - y|_T^3 \le c(|x - x_o|_T^3 + |y - x_o|_T^3),$$

and it suffices to show the following

Lemma 4.3:   Suppose $\gamma \in C^r$, $r \ge 2$, and let x_o be a critical point of E in M. Then for any $p \in [1, \infty[$ there exists a constant $c = c(p, x_o)$ such that there holds

$$||x - x_o||_\infty^{p+1} \le c(p, x_o)|x - x_o|_H^p$$

for any $x \in M$.

Proof: Since $\Gamma \in C^2$ Proposition 2.9 and Theorem I.5.1 imply that $x_o \in C^1$. Let $c_o = \left\|\left|\frac{d}{d\phi}x_o\right\|\right|_\infty < \infty$.

By monotonicity of $x \in M$ then for any $p \in [1, \infty[$

$$||x - x_o||_\infty^{p+1} \le (p+1)\, c_o \int\limits_0^{2\pi} |x - x_o|^p \, d\phi$$

$$\le c(p, x_o)\, |x - x_o|_H^p,$$

where the last estimate is a consequence of Sobolev's embedding $H^{1/2,2}(\mathbb{R}/2\pi) \hookrightarrow L^p(\mathbb{R}/2\pi)$, $\forall\, p < \infty$.

$\square$

The proof of Lemma 4.2 may now be completed by choosing $p > 2$ in Lemma 4.3.

$\square$

Verification of assumption (3.3) requires some regularity results which will be established in Chapter 5 and 6.

Lemma 4.4:   Suppose $\gamma \in C^r$, $r \geq 5$, and let $x_o \in M$ be critical for E on M, $H = H_+ \oplus H_o \oplus H_-$ the standard decomposition of H at x_o. Then H_o, H_- are finite dimensional and regular: $H_o \oplus H_- \subset C^1(\mathbb{R}/2\pi)$.

Cp. Proposition 5.6.

Recall our representation

$$T_{\mathrm{id}}G = span\{1, \sin\phi, \cos\phi\}$$

of the tangent space to the conformal group G at id , acting on $\partial B \cong \mathbb{R}/2\pi$. By conformal invariance of D for any $g \in G$ any $x \in M$, $X = X(x) \in \mathcal{C}(\Gamma)$ we have:

$$E(x \circ g) = D(X \circ g) = D(X) = E(x).$$

Hence at any critical point $x_o \in M$

(4.3)
$$\left\langle dE(x), \frac{d}{d\phi}x \cdot \xi \right\rangle = 0, \ \forall\, \xi \in T_{\mathrm{id}}G ,$$

$$d^2E(x)\left(\frac{d}{d\phi}x \cdot \xi, \eta\right) = 0, \ \forall\, \xi \in T_{\mathrm{id}}G, \forall\, \eta \in H ,$$

and the non-degeneracy condition of Theorem 3.6 cannot hold for E on any space where the conformal group G is acting.

We may attempt to normalize admissible functions x by a three-point-condition $x\left(\frac{2\pi k}{3}\right) = \frac{2\pi k}{3}$, $k \in \mathbb{Z}$. Thus we consider

$$M^* \subset \{id\} + T^* ,$$

where

$$T^* = \{\xi \in T \mid \xi\left(\frac{2\pi k}{3}\right) = 0, \ k \in \mathbb{Z}\} .$$

However, since $H^{1/2,2}$ fails to embed continuously into C^0 , taking the closure of T^* in $H^{1/2,2}$ we reobtain the full space H and the problem of degeneracy remains.

For this reason we choose to replace the pointwise constraints by integral constraints and work in the class

$$M^\dagger \subset \{id\} + T^\dagger \subset \{id\} + H^\dagger ,$$

instead, where

$$H^\dagger = \{\xi \in H \mid \int\limits_0^{2\pi} \xi\, \eta\, d\phi = 0, \ \forall\, \eta \in T_{id}G\} .$$

Clearly, assumption (3.1) will be satisfied for the triple $(M^\dagger, T^\dagger, H^\dagger)$. Moreover, by Proposition 2.9 any critical point $x_o \in M^\dagger$ of E will also be critical for E on M; by Lemma 4.2 therefore $d^2E(x_o)$ extends to $H^\dagger \times H^\dagger \subset H \times H$ and assumption (3.2) will be satisfied.

Lemma 4.5:   Suppose $\gamma \in C^r$, $r > 2$, and let $x_o \in M^\dagger$ be a non-degenerate critical point of E on $M^\dagger$. Then x_o is a C^2–diffeomorphism of the interval $[0, 2\pi]$ onto itself.

Proof: A boundary branch point gives rise to a "forced Jacobi field" $\xi \in$ $\ker d^2 E(x_0) \subset H^\dagger$; cp. Böhme-Tromba [1,Appendix I] .

$\square$

By Lemmata 4.4, 4.5 now also assumption (3.3) will be satisfied. The Palais - Smale condition is a consequence of Lemma 2.10. We summarize:

Theorem 4.6:   Suppose $\gamma \in C^r(\partial B; I\!R^n)$, $r \geq 5$, is a diffeomorphism onto a Jordan curve Γ, and assume that Γ bounds only minimal surfaces $X_o = X(x_o)$ whose normalized parametrizations $x_o \in M^\dagger$ correspond to non-degenerate critical points of E on $M^\dagger \subset id + T^\dagger \subset id + H^\dagger$ in the sense of Definition 3.2. Then the Morse inequalities (3.4) hold.

Remark 4.7:   Below we shall see that as a consequence of the "index theorem" of Böhme and Tromba [1] the non-degeneracy condition is fulfilled for almost every Γ in $I\!R^n$, $n \geq 4$, cf. Corollary 6.14.

To give an "intrinsic" characterization of non-degeneracy of a minimal surface $X_o = X(x_o) \in C_o(\Gamma)$ let us introduce the space

$$\hat{H} = \left\{ \hat{\xi} \in H^{1,2}(B; I\!R^n) | \Delta \hat{\xi} = 0, \; \hat{\xi} \text{ is proportional to } \frac{\partial}{\partial \phi}\gamma(x_o) \text{ along } \partial B \right\}$$

of harmonic surfaces tangent to X_o. Formally, $\hat{H}$ is the "tangent space" to $C_o(\Gamma)$ at X_o in $H^{1,2}(B; I\!R^3)$.

Note that since γ is a C^2–diffeomorphism onto Γ and since $x_o \in C^1$ (on account of our regularity result) the linear map (4.2)

$$dX(x_o) : \xi \mapsto h\left(\frac{d}{d\phi}\gamma(x_o) \cdot \xi \right)$$

is an isomorphism between H and $\hat{H}$.

Now compute the second variation of D on $C_o(\Gamma)$:

Let $\xi, \eta \in H$, $\hat{\xi} = dX(x_o) \cdot \xi$, $\hat{\eta} = dX(x_o) \cdot \eta \in \hat{H}$. Then by (4.1)

$$d^2 D(X_o)(\hat{\xi}, \hat{\eta}) = d^2 E(x_o)(\xi, \eta)$$

$$= \int_{\partial B} \partial_n X_o \cdot \frac{d^2}{d\phi^2} \gamma(x_o) \cdot \xi \cdot \eta \, do + \int_B \nabla(dX(x_o) \cdot \xi) \cdot \nabla(dX(x_o) \cdot \eta) dw$$

$$= \int_{\partial B} \frac{\partial_n X_o \cdot \dfrac{d^2}{d\phi^2} X_o}{\left|\dfrac{d}{d\phi}\gamma(x_o)\right|^2 \cdot \left|\dfrac{d}{d\phi}x_o\right|^2} \left(\frac{d}{d\phi}\gamma(x_o) \cdot \xi\right) \cdot \left(\frac{d}{d\phi}\gamma(x_o) \cdot \eta\right) do$$

$$+ \int_B \nabla\hat{\xi}\nabla\hat{\eta} \, dw.$$

But the expression

$$\kappa_{X_o}(\Gamma) = \frac{\partial_n X_o \cdot \dfrac{d^2}{d\phi^2} X_o}{\left|\dfrac{d}{d\phi}X_o\right|^3} = \frac{\partial_n X_o \cdot \dfrac{d^2}{d\phi^2} X_o}{\left|\dfrac{d}{d\phi}\gamma(x_o)\right|^2 \cdot \left|\dfrac{d}{d\phi}x_o\right|^2} \cdot \frac{1}{\left|\dfrac{d}{d\phi}X_o\right|}$$

equals the geodesic curvature of $\Gamma = X_o|_{\partial B}$ in the surface X_o. Hence the above formula may be simplified

(4.4)
$$d^2 D(X_o)(\hat{\xi}, \hat{\eta}) = \int_B \nabla\hat{\xi}\nabla\hat{\eta} \, dw - \int_{\partial B} \kappa_{X_o}(\Gamma)\hat{\xi}\hat{\eta} \cdot \left|\frac{d}{d\phi}X_o\right| do$$

$$= \int_B \nabla\hat{\xi}\nabla\hat{\eta} \, dw - \int_\Gamma \kappa_{X_o}(\Gamma)\hat{\xi}\hat{\eta} \, d\Gamma, \quad \forall \, \hat{\xi}, \hat{\eta} \in \hat{H},$$

and we have obtained the following result of Böhme [1] and Tromba [1]:

Proposition 4.8:   Suppose $\Gamma \in C^r$, $r \geq 3$. Then at a minimal surface $X_o \in C(\Gamma)$ the second variation of Dirichlet's integral on $\hat{H}$ is given by (4.4).

Remark 4.9:   Since $dX(x_o) : H \to \hat{H}$ is an isomorphism it is immediate that the components of the standard decompositions

$$H = H_+ \oplus H_o \oplus H_-, \quad \hat{H} = \hat{H}_+ \oplus \hat{H}_o \oplus \hat{H}_-$$

are mapped into one another under $dX(x_o)$. Moreover, dX commutes with the conformal group action. Hence $X_o = X(x_o)$ will correspond to a non-degenerate critical point in $M^\dagger$, iff $\hat{H}_0 = dX(x_0)(T_{id}G)$ and the Morse index of x_o is given by $\dim \hat{H}_-$.

We close this section with a question posed by Tromba which is related to (4.4) and the following uniquenes result of Nitsche [3] :

Theorem 4.10:   Suppose $\Gamma \subset \mathbb{R}^3$ is an analytic Jordan curve, and assume that the total curvature of Γ: $\kappa(\Gamma) \leq 4\pi$. Then (up to conformal reparametrization) Γ bounds a unique minimal surface.

The proof uses the mountain pass lemma Theorem 1.12 and the fact that under the curvature bound $\kappa(\Gamma) \leq 4\pi$ any solution $X_o = X(x_0)$ to (1.1) -(1.3) is strictly stable in the sense that for some $\lambda > 0$:

$$(4.5) \qquad\qquad d^2 D(X_o)(\hat{\xi},\, \hat{\xi}) \geq \lambda |\hat{\xi}|^2,$$

for all $\hat{\xi} \in dX(x_o)(H^\dagger)$.

Is there a way of deriving (4.5) from (4.4) *directly* ?

5. Regularity.   In this chapter we present the proofs of Proposition 2.10 and Lemma 4.4.

Propositon 5.1:   Suppose that $\gamma \in C^r$, $r \geq 3$, and let $x \in M^\dagger$ be a critical point of E on M, satisfying the variational inequality

$$(5.1) \qquad \int_{\partial B} \partial_n X \cdot \frac{d}{d\phi}\gamma(x) \cdot (x - y)\, do \leq 0$$

for all $y \in M^\dagger$, where $X = X(x)$. Then $X \in H^{2,2}(B; \mathbb{R}^n)$.

For the proof of Proposition 5.1 we need to introduce difference quotients in angular direction:

$$\partial_h \xi(\phi) \equiv \frac{1}{h}[\xi(\phi + h) - \xi(\phi)], \quad \text{etc.}$$

and translates

$$\xi_\pm(\phi) \equiv \xi(\phi \pm h), \quad \text{etc.}$$

Note the product rule

$$\partial_h(\xi\eta) = \partial_h\xi\eta_+ + \xi\partial_h\eta$$

and the following formula for integrating by parts

$$\int_0^{2\pi} \partial_h \xi\eta\, d\phi = \frac{1}{h}\int_0^{2\pi}[\xi_+\eta - \xi\eta]d\phi = \frac{1}{h}\int_0^{2\pi}[\xi\eta_- - \xi\eta]d\phi = -\int_0^{2\pi}\xi\partial_{-h}\eta\, d\phi,$$

for any $h \neq 0$, any $2\pi-$ periodic ξ and η.

Moreover, we need the following lemma.

Lemma 5.2:   If $X \in H^{1,2}(B; \mathbb{R}^n)$ is harmonic, and if for all $h \neq 0$ there holds

$$\int_B |\nabla\partial_h X|^2 dw \leq c < \infty$$

uniformly in h, then $X \in H^{2,2}(B; \mathbb{R}^n)$.

Proof: By weak compactness of L^2, there exists $f \in L^2$ such that

$$\nabla\partial_h X \xrightarrow{w} f$$

weakly in L^2 for some sequence $h \to 0$. On the other hand, for any $\varphi \in C_o^\infty(B; \mathbb{R}^n)$

$$\int_B \nabla\partial_h X \cdot \varphi dw = -\int_B \nabla X \partial_{-h}\varphi dw \to -\int_B \nabla X \frac{\partial}{\partial\phi}\varphi dw$$

as $h \to 0$, whence

$$f = \nabla \frac{\partial}{\partial \phi} X \in L^2(B).$$

But then by harmonicity of X, on $B \setminus B_{1/2}$ we can estimate

$$4^{-1} |\nabla^2 X|^2 \le |\Delta X|^2 + \frac{1}{r^2}|\nabla \frac{\partial}{\partial \phi} X|^2 \le 4|\nabla \frac{\partial}{\partial \phi} X|^2,$$

which is integrable over the annulus $B \setminus B_{1/2}(0)$. Since X is analytic in B, it follows that $\nabla^2 X \in L^2(B)$, i.e. $X \in H^{2,2}(B; \mathbb{R}^n)$.

$\square$

Proof of Proposition 5.1:   In (5.1) choose

$$y = x + \epsilon \partial_{-h} \partial_h \, x = x + \frac{\epsilon}{h^2}[x_+ - 2x + x_-]$$

$$= \left(1 - \frac{2\epsilon}{h^2}\right) x + \frac{2\epsilon}{h^2}\left(\frac{x_+ + x_-}{2}\right).$$

Note that $x_+, x_- \in M$, whence by convexity of M we have $y \in M$ if $0 \le 2\epsilon \le h^2$.

Moreover, since $x \in M^\dagger$ also $y \in M^\dagger$. Indeed, denote $\xi = x - id$; then by periodicity

$$\int_0^{2\pi} \xi_\pm \, d\phi = \int_0^{2\pi} \xi \, d\phi = 0 \, ,$$

$$\int_0^{2\pi} \xi_\pm \, \sin \phi \, d\phi = \int_0^{2\pi} \xi \, \sin(\phi \mp h) \, d\phi$$

$$= \cos h \int_0^{2\pi} \xi \, \sin \phi \, d\phi \mp \sin h \int_0^{2\pi} \xi \, \cos \phi \, d\phi = 0 \, ,$$

and likewise

$$\int_0^{2\pi} \xi_\pm \, \cos \phi \, d\phi = 0 \, .$$

I.e. $y - id = \left(1 - \frac{2\epsilon}{h^2}\right)\xi + \frac{\epsilon}{h^2}\xi_+ + \frac{\epsilon}{h^2}\xi_- \in T^\dagger$. From (5.1) we then obtain that

$$(5.2) \qquad \int_{\partial B} \partial_n X \frac{d}{d\phi}\gamma(x) \cdot \partial_{-h}\partial_h \, x \, do \ge 0.$$

On the other hand, on ∂B

$$(5.3) \quad \partial_h X = \partial_h(\gamma \circ x) = \frac{1}{h}\int_x^{x_+} \frac{d}{d\phi}\gamma(x')dx' = \frac{d}{d\phi}\gamma(x)\partial_h x + \frac{1}{h}\int_x^{x_+}\int_x^{x'} \frac{d^2}{d\phi^2}\gamma(x'')dx''dx',$$

and therefore by the product rule

(5.4)
$$\partial_{-h}\partial_h X = \frac{d}{d\phi}\gamma(x)\partial_{-h}\partial_h x +$$

$$- \left(\frac{1}{h} \int_x^{x^-} \frac{d^2}{d\phi^2}\gamma(x')dx' \right) \cdot \partial_h x_-$$

$$+ \partial_{-h} \left(\frac{1}{h} \int_x^{x_+} \int_x^{x'} \frac{d^2}{d\phi^2}\gamma(x'')dx''dx' \right).$$

Now extend $\partial_h x$, etc. to B harmonically.* Integrating by parts in (5.2) we hence infer that:

$$\int_B |\nabla \partial_h X|^2 dw \le \int_B \nabla X \ \nabla \left(\left(\frac{1}{h} \int_x^{x^-} \frac{d^2}{d\phi^2}\gamma(x')dx' \right) \partial_h x_- \right) dw$$

$$+ \int_B \nabla \partial_h X \ \nabla \left(\frac{1}{h} \int_x^{x_+} \int_x^{x'} \frac{d^2}{d\phi^2}\gamma(x'')dx''dx' \right) dw$$

$$\le c(\epsilon) \cdot \int_B \left(|\nabla X|^2 + |\nabla x|^2 \right) \left(1 + |\partial_h x|^2 \right) \ dw + \epsilon \int_B |\nabla \partial_h x|^2$$

$$+ |\nabla \partial_h X|^2 dw.$$

But on ∂B, by (5.3)

$$\partial_h x = \frac{\dfrac{d}{d\phi}\gamma(x) \cdot \partial_h X - \dfrac{1}{h} \int_x^{x_+} \int_x^{x'} \dfrac{d}{d\phi}\gamma(x) \cdot \dfrac{d^2}{d\phi^2}\gamma(x'')dx''dx'}{\left| \dfrac{d}{d\phi}\gamma(x) \right|^2}.$$

(Note that since γ is a diffeomorphism the denominator in this expression is uniformly bounded away from 0.) In consequence $D(\partial_h x)$ is bounded by the Dirichlet integral of the above right hand side:

$$\int_B |\nabla \partial_h x|^2 dw \le c \int_B |\nabla x|^2 \left(|\partial_h X|^2 + |\partial_h x|^2 \right) dw$$

$$+ c||x_+ - x||_{L^\infty}^2 \int_B |\nabla \partial_h x|^2 \ dw + c \int_B |\nabla \partial_h X|^2 \ dw,$$

and for ϵ and h sufficiently small there results

$$\int_B \left(|\nabla \partial_h X|^2 + |\nabla \partial_h x|^2 \right) dw \le c \int_B \left(|\nabla X|^2 + |\nabla x|^2 \right) \left(|\partial_h X|^2 + |\partial_h x|^2 \right) dw.$$

---

* Since $\partial_h x$ is 2π-periodic we may regard $\partial_h x$ as a function on $\partial B \cong \mathbb{R}/2\pi$.

In order to bound the products on the right two more auxiliary results are needed.

The first lemma states the "self-reproducing character" of Morrey spaces, cp. Morrey [1, Lemma 5.4.1, p.144]:

Lemma 5.3: Suppose $\varphi \in H_o^{1.2}(B)$ and $\psi \in L^1(B)$ satisfies the Morrey growth condition

$$\int\limits_{B_r(w_o) \cap B} |\psi| dw \leq c_o r^\mu$$

for all $r > 0, w_o \in B$ with uniform constants c_o and $\mu > 0$. Then $\psi \varphi^2 \in L^1(B)$ and for all $r > 0, w_o \in B$ there holds

$$\int\limits_{B_r(w_o) \cap B} |\psi \varphi^2| dw \leq c_1 c_o\, r^{\mu/2} \int\limits_B |\nabla \varphi|^2 dw$$

with a uniform constant c_1.

The second auxiliary result establishes the Morrey growth condition for the functions $\psi = |\nabla X|^2 + |\nabla x|^2$.

Lemma 5.4:   Under the assumptions of Proposition 5.1 there exist constants $c_o, \mu > 0$ such that for all $r > 0$, $w_o \in B$ there holds

$$\int\limits_{B_r(w_o) \cap B} |\nabla X|^2 + |\nabla x|^2 dw \leq c_o r^\mu \int\limits_B |\nabla X|^2 + |\nabla x|^2 dw.$$

Proof: Fix $w_o = e^{i\phi_o} \in \partial B$, and let x_o be the mean of x over the "annulus" $(B_{2r}(w_o) \backslash B_r(w_o)) \cap \partial B$; also let $\tau \in C^\infty$ be a non-increasing function of the distance $|w - w_o|$ satisfying the conditions $0 \leq \tau \leq 1$, $\tau \equiv 1$ if $|w - w_o| \leq 2r$, $\tau \equiv 0$ if $|w - w_o| \geq 3r$, $|\nabla \tau| \leq c/r$, $|\nabla^2 \tau| \leq c/r^2$.

Then for $3r < 1$ the function

$$y = x - \tau^2(x - x_o) = (1 - \tau^2)x + \tau^2 x_o \in M.$$

Indeed, a.e. on $\partial B \hat{=} [0, 2\pi]$ we have

$$\frac{d}{d\phi} y = (1 - \tau^2)\frac{d}{d\phi} x - \frac{d}{d\phi}(\tau^2)(x - x_o)$$

and $\frac{d}{d\phi} x \geq 0$ a.e. while by monotonicity of x and our choice of τ also the second term is non-negative.
Now pretend that inequality (5.1) is valid for y. Then we obtain

$$\int\limits_{\partial B} \partial_n X \frac{d}{d\phi} \gamma(x)(x - x_o)\tau^2 do \leq 0.$$

Let $X_o = \gamma(x_o) \in \Gamma$. Then on ∂B

$$(5.5) \qquad X - X_o = \frac{d}{d\phi}\gamma(x)(x - x_o) - \int\limits_{x_o}^{x}\int\limits_{x'}^{x} \frac{d^2}{d\phi^2}\gamma(x'')dx''dx'.$$

Upon an integration by parts we hence obtain that for any pre - assigned $\delta > 0$ with a constant c_δ there holds:

$$\int\limits_{B} |\nabla X|^2 \tau^2 dw \leq \int\limits_{\partial B} \partial_n X(X - X_o)\tau^2 do + 2\int\limits_{B} |\nabla X||X - X_o||\nabla\tau|\tau dw$$

$$\leq -\int\limits_{\partial B} \partial_n X \left(\int\limits_{x_o}^{x}\int\limits_{x'}^{x} \frac{d^2}{d\phi^2}\gamma(x'')dx''dx'\right)\tau^2 do +$$

$$+ \delta\int\limits_{B} |\nabla X|^2 \tau^2 dw + c_\delta\int\limits_{B} |\nabla\tau|^2|X - X_o|^2 dw.$$

After another integration by parts

$$-\int\limits_{\partial B} \partial_n X \left(\int\limits_{x_o}^{x}\int\limits_{x'}^{x} \frac{d^2}{d\phi^2}\gamma(x'')dx''dx'\right)\tau^2 do$$

$$\leq c\cdot\int\limits_{B} |\nabla X||\nabla x||x - x_o|\tau^2 + |\nabla X|\,|\nabla\tau||x - x_o|^2\tau dw$$

$$\leq c\|x - x_o\|_{L^\infty(B_{3r}(w_0))}\int\limits_{B} \left(|\nabla X|^2 + |\nabla x|^2\right)\tau^2 + |x - x_o|^2|\nabla\tau|^2 dw.$$

Thus, for $0 < r < r_o$ sufficiently small (depending on δ and the modulus of continuity of x) we obtain the estimate

$$(5.6) \qquad \int\limits_{B} |\nabla X|^2 \tau^2 dw \leq \delta\int\limits_{B} \left(|\nabla X|^2 + |\nabla x|^2\right)\tau^2 dw$$

$$+ c\int\limits_{B} \left(|X - X_o|^2 + |x - x_o|^2\right)|\nabla\tau|^2 dw.$$

Now extend x harmonically to $B \cap B_{3r}(w_0)$. Note that

$$(\nabla x)\tau = \nabla((x - x_o)\tau) - (x - x_o)\nabla\tau$$

and that

$$\Delta((x - x_o)\tau) = 2\nabla x\,\nabla\tau + (x - x_o)\Delta\tau =: f$$

has support only on $B_{3r}(w_0)\backslash B_{2r}(w_o)$. Let

$$(5.7) \quad (x - x_o)\tau = \frac{\dfrac{d}{d\phi}\gamma(x)\cdot(X - X_o) + \int\limits_{x_o}^{x}\int\limits_{x'}^{x}\dfrac{d}{d\phi}\gamma(x)\cdot\dfrac{d^2}{d\phi^2}\gamma(x'')dx''dx'}{\left|\dfrac{d}{d\phi}\gamma(x)\right|^2}\cdot\tau =: \varphi_o$$

on $\partial B \cup (B \cap \partial B_{3r}(w_o))$. Recall the variational characterization

$$D(\varphi) \le D(\psi) - \int_{B \cap B_{3r}(w_0)} f(\varphi - \psi) dw, \quad \forall \psi \in \varphi_o + H_o^{1,2}(B \cap B_{3r}(w_o))$$

for the solution $\varphi = (x - x_o)\tau$ of the equation $\triangle \varphi = f$ in $B \cap B_{3r}(w_o)$ with boundary data φ_o. Upon inserting $\psi = \varphi_o$ this implies that

$$\int_B |\nabla x|^2 \tau^2 dw \le c \int_B |\nabla X|^2 \tau^2 dw + c \int_B \left(|X - X_o|^2 + |x - x_o|^2 \right) \left(|\nabla \tau|^2 + |\triangle \tau| \right) dw$$

$$+ c || \, |X - X_o| + |x - x_o| \, ||_{L^\infty (B \cap B_{3r}(w_o))}^2 \int_B \left(|\nabla \tau|^2 + |\nabla x|^2 \tau^2 \right) \, dw.$$

Together with (5.6) this estimate (for sufficiently small $r > 0$, depending on the modulus of continuity of x) yields that

$$\int_{B_r(w_o) \cap B} \left(|\nabla X|^2 + |\nabla x|^2 \right) \, dw \le cr^{-2} \int_{(B \cap B_{3r}(w_o) \setminus B_{2r}(w_o))} \left(|X - X_o|^2 + |x - x_o|^2 \right) dw.$$

The right hand side in this expression can now be further estimated by means of a Poincaré - Sobolev - type inequality, cp. Lemma 5.5 below: (We omit writing w_o in the sequel.)

$$\int_{B \cap (B_{3r} \setminus B_r)} \left(|X - X_o|^2 + |x - x_o|^2 \right) dw \le cr^2 \int_{B \cap (B_{3r} \setminus B_r)} \left(|\nabla X|^2 + |\nabla x|^2 \right) dw +$$

$$+ c \left(\int_{\partial B \cap B_{2r} \setminus B_r} (X - X_o) \, do \right)^2$$

$$+ \left(\int_{\partial B \cap B_{2r} \setminus B_r} (x - x_o) \, do \right)^2 .$$

By choice of x_o

$$\int_{\partial B \cap B_{2r} \setminus B_r} (x - x_o) \, do = 0,$$

while by (5.5) and an application of the trace theorem:

$$\int_{\partial B \cap B_{2r} \setminus B_r} (X - X_o)\, do = \int_{\partial B \cap B_{2r} \setminus B_r} \frac{d}{d\phi}\gamma(x_o)(x - x_o)\, do +$$

$$+ \int_{\partial B \cap B_{2r} \setminus B_r} \int_{x_o}^{x} \int_{x_o}^{x'} \frac{d^2}{d\phi^2}\gamma(x'')\, dx''\, dx'\, do$$

$$\leq c \cdot \int_{\partial B \cap B_{2r} \setminus B_r} |x - x_o|^2\, do$$

$$\leq cr \int_{B \cap B_{2r} \setminus B_r} |\nabla x|^2\, dw + c \left(\int_{\partial B \cap B_{2r} \setminus B_r} (x - x_o)\, do \right)^2$$

$$= cr \int_{B \cap B_{2r} \setminus B_r} |\nabla x|^2\, dw.$$

I.e. if we let

$$\Phi(r) = \int_{B \cap B_r(w_o)} \left(|\nabla X|^2 + |\nabla x|^2 \right)\, dw,$$

we obtain the difference inequality

$$\Phi(r) \leq c_1[\Phi(3r) - \Phi(r)].$$

We now use Widman's [1] "hole filling" trick: Add $c_1 \Phi(r)$ to this inequality, and divide by $1 + c_1$ to obtain

$$\Phi(r) \leq \frac{c_1}{1 + c_1}\Phi(3r).$$

Let $\Theta = \frac{c_1}{1 + c_1}$. By iteration, for any $r > 0$, $k \in I\!N$ such that $3^k r \leq r_o$ there results

$$\Phi(r) \leq \Theta^k \Phi(3^k r).$$

Finally, for any $r \in\,]0, r_o[$ there is exactly one $k \in I\!N$ such that

$$r_o \in\,]3^{k-1}r,\, 3^k r].$$

Clearly, k is the smallest integer

$$k > \ln\left(\frac{r_o}{r}\right) \Big/ \ln 3,$$

whence

$$\Phi(r) \leq \Theta^{-1}\left(\frac{r_o}{r}\right)^{\ln \Theta / \ln 3} \Phi(r_o).$$

We may now choose

$$\mu = -\ln\Theta / \ln 3 > 0,$$

$$c_o = \frac{\Phi(r_o)}{\Theta r_o^\mu}$$

to complete the proof of the lemma, if $y \in M^\dagger$.

To achieve the normalization $y \in M^\dagger$ we modify x near three points $w_j = e^{i\phi_j}$, $j = 1, 2, 3$ simultaneously where the angles ϕ_j differ by approximately $\frac{2\pi}{3}$. Then, if we let

$$\Phi(r) = \sup_{w_0 \in \partial B} \int_{B \cap B_r(w_0)} (|\nabla X|^2 + |\nabla x|^2)\, dw$$

our above reasoning leads us to an estimate

$$\Phi(r) \leq \Theta\, \Phi(3r)$$

with a fixed constant $\Theta < 1$ for all $r \leq r_0$, and the desired conclusion follows by iteration as before.

□

Proof of Proposition 5.1   (completed):   Extend X to $\mathbb{R}^2$ by reflection

$$X(w) = X\left(\frac{w}{|w|^2}\right), \quad \forall\, w \notin B.$$

Choose $r \in\,]0, 1/2\,]$ and let $\tau \in C_o^\infty(B_{2r}(0))$ satisfy $\tau \equiv 1$ on $B_r(0)$. Cover B by balls of radius r in such a way that at any point $w \in B$ at most k balls of the cover intersect, k independent of r. Let the balls B_i of this cover be centered at points w_i, and let $\tau_i(w) \equiv \tau(w - w_i)$. Then

$$c^{-1} \int_B \left(|\nabla \partial_h X|^2 + |\nabla \partial_h x|^2\right) dw \leq \sum_i \int_{\mathbb{R}^2} \left(|\nabla X|^2 + |\nabla x|^2\right) \left(|\partial_h X|^2 + |\partial_h x|^2\right) \tau_i^2\, dw.$$

Moreover, by Lemma 5.4 the function $\psi = \left(|\nabla X|^2 + |\nabla x|^2\right)$ satisfies a Morrey growth condition as in Lemma 5.3. Applying Lemma 5.3 with this function ψ and $\varphi_i = \partial_h X \tau_i$, resp. $\varphi_i = \partial_h x \tau_i \in H_o^{1,2}(B_{2r}(w_i))$ we obtain that

$$\int_{B_{2r}(w_i)} \left(|\nabla X|^2 + |\nabla x|^2\right) \left(|\partial_h X|^2 + |\partial_h x|^2\right) \tau_i^2\, dw$$

$$\leq c r^{\mu/2} \int_{B_2(0)} \left(|\nabla X|^2 + |\nabla x|^2\right) dw \int_{B_{2r}(w_i)} \left(|\nabla \partial_h X|^2 + |\nabla \partial_h x|^2\right) dw$$

$$+ c r^{\mu/2} \int_{B_2(0)} |\nabla X|^2 + |\nabla x|^2) dw \int_{B_{2r}(w_i)} \left(|\partial_h X|^2 + |\partial_h x|^2\right) |\nabla \tau_i|^2 dw.$$

Summing over i we may infer that with a constant c_1 independent of r there holds:

$$\int_B \left(|\nabla \partial_h X|^2 + |\nabla \partial_h x|^2\right) dw \leq c_1 r^{\mu/2} \int_{B_2(0)} \left(|\nabla \partial_h X|^2 + |\nabla \partial_h x|^2\right) dw +$$

$$+ c r^{-2} \int_{B_2(0)} \left(|\partial_h X|^2 + |\partial_h x|^2\right) dw.$$

However, bounds for X and its derivatives on B imply similar bounds on $B_2(0)$. Moreover $\partial_h X \to \frac{d}{d\phi} X$ strongly in L^2. Hence we deduce that

$$\int\limits_B \left(|\nabla \partial_h X|^2 + |\nabla \partial_h x|^2 \right) dw \le c_1 r^{\mu/2} \int\limits_B \left(|\nabla \partial_h X|^2 + |\nabla \partial_h x|^2 \right) dw + cr^{-2}$$

and for sufficiently small $r > 0$ there results a uniform a-priori bound

$$\int\limits_B \left(|\nabla \partial_h X|^2 + |\nabla \partial_h x|^2 \right) dw \le c$$

independent of h. By Lemma 5.2 $X \in H^{2,2}(B; I\!R^n)$, and the proof is complete .

$\square$

For completeness, we give a proof of Poincare's inequality that we have used in the proof of Lemma 5.4:

Lemma 5.5:   Let $3r < 1, w_o \in \partial B$, $B_r = B_r(w_o)$,   etc.   Set $G = B \cap (B_{3r} \backslash B_r)$, $S = \partial B \cap B_{2r} \backslash B_r$. Then for any $\varphi \in H^{1,2}(G)$ there holds

$$\int\limits_G |\varphi|^2 dw \le cr^2 \int\limits_G |\nabla\varphi|^2 dw + c\left(\int\limits_S \varphi do \right)^2$$

with a constant c independent of r, w_o, and φ.

Proof: Suppose by contradiction that for a sequence $\{\varphi_m\}$ in $H^{1,2}(G)$

$$1 = \int\limits_G |\varphi_m|^2 dw \ge mr^2 \int\limits_G |\nabla\varphi_m|^2 dw + m\left(\int\limits_S \varphi_m do \right)^2 .$$

Then $\{\varphi_m\}$ is bounded in $H^{1,2}(G)$ and we may assume that $\varphi_m \xrightarrow{w} \varphi$ weakly in $H^{1,2}(G)$. By the Rellich-Kondrakov theorem also $\varphi_m \to \varphi$ strongly in $L^2(G)$ and in $L^1(S)$. Moreover, $\nabla\varphi_m \to 0$ strongly in $L^2(G)$, $\int\limits_S \varphi_m do \to 0$, whence $\varphi \equiv \text{const} = 0$. But by strong L^2-convergence we must have $\int\limits_G |\varphi|^2 dw = 1$.

The contradiction proves the estimate for fixed r. That c is in fact independent of r and w_o is seen by scaling.

$\square$

Proposition 5.6:   Suppose $\gamma \in C^r$, $r \ge 5$, and let $x_o \in M$ be a critical point of E on M corresponding to a minimal surface $X_o = X(x_o) \in$

$\mathcal{C}(\Gamma)$ spanning Γ. Let $H = H_+ \oplus H_o \oplus H_-$ be the standard decomposition of $H = H^{1/2,2}(I\!\!R/2\pi)$ induced by $d^2E(x_o)$. Then $\dim(H_o \oplus H_-) < \infty$ and $H_o \oplus H_- \subset C^1$.

Proof: Let $\hat{H} = dX(x_o)(H)$ be the space of $H^{1,2}$−vectorfields "tangent" to X_o. Recall that

(5.8)
$$d^2 D(X_o)(\hat{\xi}, \hat{\eta}) = \int_B \nabla\hat{\xi}\nabla\hat{\eta}\,dw + \int_{\partial B} \kappa_{X_o}(\Gamma)\hat{\xi}\hat{\eta}\left|\frac{d}{d\phi}X_o\right|do$$
$$= d^2 E(x_o)(\xi,\eta)$$

for all $\xi,\eta \in H, \hat{\xi} = dX(x_o).\xi, \hat{\eta} = dX(x_o)\cdot\eta \in \hat{H}$. $d^2D(X_o)$ induces a splitting $\hat{H} = \hat{H}_+ \oplus \hat{H}_o \oplus \hat{H}_-$, and since $dX(x_o) = h\left(\frac{d}{d\phi}\gamma(x_o).\right)$ is a topological isomorphism of H onto $\hat{H}$ from (5.8) it is clear that also $dX(x_o)|_{H_+} : H_+ \to \hat{H}_+$, $dX(x_o)|_{H_o} : H_o \to \hat{H}_o$, and $dX(x_o)|_{H_-} : H_- \to \hat{H}_-$ are topological isomorphisms.

It hence suffices to show that $\dim\left(\hat{H}_o + \hat{H}_-\right) < \infty$. But this follows from the Gårding type inequality

(5.9)
$$d^2 D(X_o)(\hat{\xi},\hat{\xi}) \geq \int_B |\nabla\hat{\xi}|^2 dw - c\int_{\partial B} |\hat{\xi}|^2 do$$

and a standard argument: If $\{\hat{\xi}_m\}$ is an orthonormal system in $\hat{H}_o \oplus \hat{H}_-$, (5.9) implies that

(5.10)
$$1 = \int_B |\hat{\xi}_m|^2 + |\nabla\hat{\xi}_m|^2 dw \leq c\int_{\partial B} |\hat{\xi}_m|^2 do + \int_B |\hat{\xi}_m|^2 dw$$

for all m. But the embeddings $H^{1,2}(B;I\!\!R^n) \hookrightarrow L^2(B;I\!\!R^n)$ resp. $\hookrightarrow L^2(\partial B;I\!\!R)$ are compact. Hence $\{\hat{\xi}_m\}$ must be finite, or (5.10) would eventually be violated for a sequence $\{\hat{\xi}_m\}$ weakly tending to 0 in $H^{1,2}(B;I\!\!R^n)$.

To obtain regularity of an element $\xi \in H_o \oplus H_-$, we once again use difference quotients $\partial_{-h}\partial_h\xi$, etc., as introduced at the beginning of this chapter.

(5.11)
$$\partial_h\hat{\xi} = \partial_h(dX(x_o))\cdot\xi_+ + dX(x_o)\cdot\partial_h\xi$$

and

$$\partial_{-h}\partial_h\hat{\xi} = \partial_{-h}\left(\partial_h(dX(x_o))\cdot\xi_+\right) + \partial_{-h}\left(dX(x_o)\right)\cdot(\partial_h\xi)_- + dX(x_o)\cdot\partial_{-h}\partial_h\xi.$$

Extend $\xi, \partial_h\xi$, etc. harmonically into B. (Since ξ is 2π-periodic, we may think of ξ as a member of $H^{1/2,2}(\partial B)$, as well.) Note that in the

$H^{1/2,2}$–scalar product induced by the restriction $H^{1,2}(B; I\!R^n) \ni X \to X|_{\partial B} \in H^{1/2,2}(\partial B; I\!R^n)$ *

$$(\xi, \eta)_{1/2} = \int_B \nabla \xi \, \nabla \eta \, dw$$

where ξ, η denote the harmonic extensions of ξ, η to B, we have

(5.12) $(\xi, \, \partial_{-h}\partial_h\xi)_{1/2} = -|\partial_h\xi|^2_{1/2} \le 0.$

Hence, if $\xi \in H_- \oplus H_o$ it follows that $d^2 E(x_o)(\xi, \, \partial_{-h}\partial_h\xi) \ge 0$. Therefore

(5.13) $d^2 D(x_o)(\hat\xi, \partial_{-h}\partial_h\hat\xi) = d^2 E(x_o)(\xi, \partial_{-h}\partial_h\xi) + R_1 \ge R_1$

where by Sobolev's embedding theorem and Young's inequality

$$R_1 = d^2 D(X_o) \left(\hat\xi, \partial_{-h}(\partial_h(dX(x_o)) \cdot \xi_+) + \partial_{-h}(dX(x_o)) \cdot (\partial_h\xi)_- \right)$$

$$\le c \cdot \int_B |\nabla\partial_h\hat\xi| \left(|\xi_+| + |\nabla\xi_+|\right) + |\nabla\hat\xi| \left(|\partial_h\xi_-| + |\nabla\partial_h\xi_-|\right) dw$$

$$+ c \int_{\partial B} |\partial_h\hat\xi| \, |\xi_+| + |\hat\xi| \, |\xi_+| + |\hat\xi| \, |\partial_h\xi_-| do$$

$$\le \epsilon \int_B |\nabla\partial_h\hat\xi|^2 + |\nabla\partial_h\xi|^2 dw + C(\epsilon).$$

$C(\epsilon)$ may depend on the $H^{1,2}$–norm of $\hat\xi$, resp. the $H^{1/2,2}$–norm of ξ, and ϵ. The estimate for R_1 requires C^3–estimates for X_o.

But by (5.8)

(5.14) $d^2 D(X_o)(\hat\xi, \partial_{-h}\partial_h\hat\xi) = - \int_B |\nabla\partial_h\hat\xi|^2 dw - R_2$

where

$$R_2 = \int_{\partial B} \kappa_{X_o}(\Gamma)|\partial_h\hat\xi|^2 \left|\frac{d}{d\phi}X_o\right| do + \int_{\partial B} \partial_h \left(\kappa_{X_o}(\Gamma)\left|\frac{d}{d\phi}X_o\right|\right) \hat\xi \cdot \partial_h\hat\xi dw$$

$$\le c \int_{\partial B} |\partial_h\hat\xi|^2 do + c \le \epsilon \int_B |\nabla\partial_h\hat\xi|^2 dw + C(\epsilon).$$

Again we require $X_o \in C^3(\overline{B}; I\!R^n)$.

Moreover, note that by (5.11)

$$\int_B |\nabla\partial_h\xi|^2 dw \le c \int_B |\nabla\partial_h\hat\xi|^2 dw + C.$$

---

* cp. the proof of Lemma 2.5

Finally, combining (5.13), (5.14)

$$\int\limits_B |\nabla \partial_h \hat{\xi}|^2 dw \leq |R_1| + |R_2| \leq \epsilon \int\limits_B |\nabla \partial_h \hat{\xi}|^2 dw + C(\epsilon),$$

and it follows that

$$\int\limits_B |\nabla \partial_h \hat{\xi}|^2 dw \leq C(\epsilon)$$

uniformly in $h > 0$. By Lemma 5.2 $\hat{\xi} \in H^{2,2}(B; I\!\!R^n)$.

Inserting fourth order difference quotients $\partial_{-h}\partial_h\partial_{-h}\partial_h\xi$, in a similar manner for $X_o \in C^4(\overline{B}; I\!\!R^3)$ we obtain that $\hat{\xi} \in H^{3,2}((B; I\!\!R^n)) \hookrightarrow C^1(\overline{B}; I\!\!R^n)$, and hence the claim.

$\Box$

6. Historical remarks.   The solution of Plateau's problem fell into a period of very active research in variational problems. Only a few years before Jesse Douglas' and Tibor Radó's work on minimal surfaces L. Ljusternik and L. Schnirelmann had developed powerful new variational methods which enabled them to establish the existence of 3 distinct closed geodesics on any compact surface of genus zero. Also in the 20's Marston Morse outlined the general concept of what is now known as Morse theory: A method for relating the number and types (minimum, saddle) of critical points of a functional to topological properties of the space over which the functional is defined. Quite naturally, Morse and his contemporaries were eager to apply this new theory to the Plateau problem.

In the following we briefly survey the Morse theorical results obtained for the Plateau problem by Morse-Tompkins [1] and independently by Shiffman [1] in 1939. Necessarily, this account cannot accurately present all the details of these approaches. Nevertheless, I hope that I have faithfully portrayed the main ideas.

The work of Morse-Tompkins and Shiffman.

Morse-Tompkins and Shiffman approach the Plateau problem in the frame set by Douglas. I.e. surfaces spanning Γ are represented as monotone maps $x \in M^*$ of the interval $[0, 2\pi]$ onto itself, preserving the points $(2\pi k)/3$, $k = 1, 2, 3$ and their areas are expressed by the Dirichlet-Douglas integral E, cf. (2.3), (2.4).

The non-compactness of the space M^* is no problem. In fact, the principles that Morse had developed apply to any functional $\mathcal{E}$ on any metric space $(\mathcal{M}, d)$ provided the conditions of "regularity at infinity", "weak upper-reducibility", and "bounded compactness" are satisfied. This latter condition is crucial. It requires that for any $\alpha \in I\!R$ the set

$$(6.1) \qquad \overline{\mathcal{M}}_\alpha = \{x \in \mathcal{M} \mid \mathcal{E}(x) \le \alpha\}$$

is compact.

By Proposition 2.1, the functional $\mathcal{E} = E$ will satisfy the condition of bounded compactness on $\mathcal{M} = M^*$ if we endow M^* with the C^o−topology of uniform convergence.

This choice of topology therefore is the natural choice that Morse-Tompkins and Shiffman take. However, in this topolgy E is only lower semi-continuous on M^*, cf. Remark I.3.2.

For a functional which is not differentiable the notions of a critical point and its critical type are defined with reference to neighborhoods of a point $x_o \in \mathcal{M}$ with $\mathcal{E}(x_o) = \beta$ in the level set $\overline{\mathcal{M}}_\beta$.

Definition 6.1:   Let $\beta \in I\!R$, and let $U \subset \overline{\mathcal{M}}_\beta$ be relatively open, $\varphi :$ $U \times [0, 1] \to \mathcal{M}$ a continuous deformation such that $\varphi(\cdot, 0) = \text{id}|_U$. Let $V \subset\subset$

U. φ possesses a *displacement function* $\delta : I\!R^+ \cup \{0\} \to I\!R^+ \cup \{0\}$ on V iff for all $x \in V$, $0 \le s \le t \le 1$ there holds

$$\mathcal{E}(\varphi(x,s)) - \mathcal{E}(\varphi(x,t)) \ge \delta\left(d(\varphi(x,s),\varphi(x,t))\right)$$

and

$$\delta(e) = 0 \quad \text{iff } e = 0.$$

The deformation φ is an $\mathcal{E}-deformation$ on U if φ possesses a displacement function on any $V \subset\subset U$.

Definition 6.2: $x_o \in \mathcal{M}$ with $\mathcal{E}(x_o) = \beta$ is *homotopically regular* if there is a neighborhood U of x_o in $\overline{\mathcal{M}}_\beta$ and an $\mathcal{E}-$deformation φ on U which displaces x_o (in the sense that $\varphi(x_o,1) \ne x_o$). Otherwise, x_o is *homotopically critical*.

Remark 6.3: Definitions 6.1, 6.2 imply that for a homotopically regular point x_o there exists a deformation $\varphi : U \times [0,1] \to \mathcal{M}$ of a neighborhood U of x_o in $\overline{\mathcal{M}}_\beta$ such that

i) $\qquad \varphi(x,0) = x, \quad \forall x \in U,$

ii) $\qquad \mathcal{E}(\varphi(x,t))$ is non-increasing in t, $\forall x \in U,$

iii) $\qquad$ For any $V \subset\subset U$ there is a number $\epsilon > 0$ such that

$$\varphi(V,1) \subset \mathcal{M}_{\beta-\epsilon}.$$

By (6.1) at a regular value β finitely many such neighborhoods cover

$$\{x \in \mathcal{M} | \mathcal{E}(x) = \beta\}.$$

Piecing deformations together we thus obtain a homotopy equivalence

$$\overline{\mathcal{M}}_\beta \simeq \overline{\mathcal{M}}_{\beta-\epsilon}$$

for some $\epsilon > 0$, for any regular value β, as in the differentiable case.

Examples 6.4: $\quad$ i) $\quad$ If x_o is a relative minimum of $\mathcal{E}$ on $\mathcal{M}$ then x_o is homotopically critical. Indeed, for suitable $U \ni x_o$ we have $U \cap \overline{\mathcal{M}}_\beta = \{x_o\}$, and U cannot admit an $\mathcal{E}-$deformation which displaces x_o.

ii) $\quad$ Let $\mathcal{M} = I\!R^2, \mathcal{E}(x,y) = x^2 - y^2$. The point $(0,0)$ is homotopically critical since $\overline{\mathcal{M}}_o$ is connected while for any $\epsilon > 0$ and any neighborhood U of $(0,0)$ the set $\overline{\mathcal{M}}_{-\epsilon} \cap U$ is not.

iii) $\quad$ Let $\mathcal{M} = I\!R$, $\mathcal{E}(x) = x^d, d \in I\!N$. The point $x_o = 0$ is homotopically critical iff d is even.

Examples 6.4 illustrate that the concept of a homotopically critical point is natural but somewhat delicate. In general, in order to be able to decide whether for a differentiable functional $\mathcal{E} \in C^1(\mathcal{M})$ a critical point $x_o \in \mathcal{M}$ (in the sense that $d\mathcal{E}(x_o) = 0$) is also homotopically critical one needs to analyze the topology of the level set $\overline{\mathcal{M}}_\beta$ near x_o. Unless x_o is a relative minimum, this analysis in general requires that $\mathcal{E} \in C^2$ near x_o and that $d^2\mathcal{E}(x_o)$ is non-degenerate.

For the Plateau problem we have the following result, Morse - Tompkins [1, Theorem 6.2]:

Lemma 6.5:   If $x_o \in M^*$ is homotopically critical for E on M^* endowed with the C^o- topology, then $X_o = X(x_o)$ parametrizes a minimal surface spanning Γ.

Information concerning the critical type of x_o is captured in the following

Definition 6.6:   Let $x_o \in \mathcal{M}$ be an isolated homotopically critical point of $\mathcal{E}$ with $\mathcal{E}(x_o) = \beta$, and let $U \subset \overline{\mathcal{M}}_\beta$ be a neighborhood of x_o containing no other homotopically critical point. Then

$$t_k(x_o) = \lim_{\alpha \nearrow \beta} \inf \, rank \left(H_k \left(U, \overline{\mathcal{M}}_\alpha \right) \right)$$

is the k-th *type number* of x_o.

The following observation is crucial:

Lemma 6.7:   The numbers $t_k(x_o)$ are independent of U.

Example 6.8.   i)   If x_o is a strict relative minimum of $\mathcal{E}$ on a metric space $\mathcal{M}$, then

$$t_k(x_o) = \begin{cases} 1, & k = 0 \\ 0, & \text{else} \end{cases} .$$

ii)   If x_o is a non-degenerate critical point of $\mathcal{E} \in C^2(\mathcal{M})$

$$t_k(x_o) = \begin{cases} 1, & k = \text{Index}(x_o) \\ 0, & \text{else} \end{cases} .$$

Unless x_o falls into the categories i), ii) of Example 6.8 in general it may be impossible to compute its type numbers.

Now let

$$R_k = rank \left(H_k \left(\mathcal{M} \right) \right),$$

$$T_k = \sum_{x \text{ hom. crit.}} t_k(x)$$

be the Betti numbers of $\mathcal{M}$ and type numbers of $\mathcal{E}$, resp. Then Morse's theory asserts:

Theorem 6.9: Suppose $\mathcal{E} : \mathcal{M} \to I\!R$ satisfies the conditions of "regularity at infinity", "weak upper-reducibility ", and "bounded compactness" and assume that $\mathcal{E}$ possesses only finitely many homotopically critical points. Then the inequalities hold:

$$T_k \geq R_k,$$

$$\sum_{k=0}^{m}(-1)^{m-k} T_k \geq \sum_{k=0}^{m}(-1)^{m-k} R_k,$$

$$\sum_{k=0}^{\infty}(-1)^k T_k = \sum_{k=0}^{\infty}(-1)^k R_k.$$

For the Plateau problem Theorem 6.9 has the following corollary, cp. Morse-Tompkins [1, Corollary 7.1], which is slightly weaker than our result Theorem 2.11:

Theorem 6.10: Suppose Γ bounds two distinct strict relative minima X_1, X_2 of D. Then there exists an unstable minimal surface X_3 spanning Γ, distinct from X_1, X_2.

Proof: Note that since M^* is contractible its Betti-numbers

$$R_k = \begin{cases} 1, & k = 0 \\ 0, & \text{else} \end{cases}.$$

By Example 6.8, i)

$$T_o \geq 2,$$

whence Theorem 6.9 for $m = 1$ gives the relation

$$T_1 \geq T_o - 1 \geq 1.$$

Hence E must possess a critical point x_3 such that any neighborhood of x_3 in M^* contains points x with $E(x) < E(x_3)$. I.e. $X_3 = X(x_3)$ is an unstable minimal surface.

$$\square$$

But what is the relation of Theorem 6.9 in the case of the Plateau problem with our Theorem 4.6? Are these results equivalent - at least in case Γ spans only finitely many minimal surfaces which are non-degenerate in the sense of Definition 3.2? In particular, is it possible to identify $T_k = C_k$ in this case? The answer to this question is unknown. In fact, the C^o-topology seems too coarse to allow us to compute the homology of C^o-neighborhoods of critical points of E in terms of the second variation of E near such points - even if we use the $H^{1/2, 2}$-expansion Lemma 4.2 . It is not even clear if such points will be homotopically critical points of E in the sense of Definition 6.2 and will register in Theorem 6.9 at all.

The technical complexity and the use of a sophisticated topological machinery (which is not shadowed in our presentation) moreover tend to make Morse-Tompkins' original paper unreadable and inaccessible for the non-specialist, cf. Hildebrandt [4, p. 324].

Confronting Morse-Tompkins' and Shiffman's approach with that given in Chapter 4 we see how much can be gained in simplicity and strength by merely replacing the C^o−topology by the $H^{1/2\,,2}$−topology and verifying the Palais - Smale - type condition stated in Lemma 2.10.

However, in 1964/65 when Palais and Smale introduced this condition in the calculus of variations it was not clear that it could be meaningful for analyzing the geometry of surfaces, cf. Hildebrandt [4, p. 323 f.].

Instead, a completely new approach was taken by Böhme and Tromba [1] to tackle the problem of understanding the global structure of the set of minimal surfaces spanning a wire.

The Index Theorem of Böhme and Tromba and its consequences.

Böhme and Tromba turn around completely our view of the classical Plateau problem. If to this moment we have only looked at surfaces with a *fixed* boundary Γ, now Böhme and Tromba consider the bundle of all surfaces spanning *any* Jordan curve in $I\!R^n$. If we had so far tried to understand the structure of minimal surfaces with *given* boundary, Böhme and Tromba analyze the structure of the set of *all* branched minimal surfaces in $I\!R^n$. The information that we need in order to solve the Plateau problem for a given wire is contained in the properties of two differentiable maps: The (bundle) projection Π of a surface to its boundary, and the conformality operator K. Without going into technicalities we now present the main ideas of Böhme' and Tromba's approach.

For details we refer the interested reader to the original paper of Böhme - Tromba [1] and to the papers by Schüffler - Tomi [1], Söllner [1], Thiel [1] ,[2] on extensions and simplifications of their approach.

Let

$$\mathcal{A}$$

be the space of diffeomorphisms $\gamma : \partial B \to I\!R^n$; this is the space of (parametrized) curves.

Let

$$D = \bigcup_{\nu \in I\!N_o} D_\nu$$

be the space of monotone parametrizations x of $\partial B = I\!R/2\pi$ where

$$D_\nu = \Big\{ x \in D | \frac{dx}{d\phi} = \prod_{k=1}^{q} (\phi - \phi_j)^{\nu_k} \Theta, \Theta > 0 \ ,$$

$$0 \leq \phi_1 < ... < \phi_q < 2\pi, \ \nu_k \in I\!N, \ \sum_{k=1}^{q} \nu_k = \nu \Big\}.$$

In order for $x \in D_\nu$ to be monotone, necessarily each ν_k must be even.

We may think of D as a subset of our space M, introduced in (2.2), and at each $x \in D_\nu$ the tangent space $T_x D_\nu \subset H$, cf. Section 4.

Finally let

$$\eta = \bigcup_{\lambda, \nu \in I\!N_o} \eta_\nu^\lambda$$

be the space of harmonic surfaces in $I\!R^n$ decomposed into bundles η_ν^λ consisting of harmonic surfaces with p interior and q boundary branch points of orders

$\lambda_1, ..., \lambda_p$ resp. $\nu_1, ..., \nu_q$ such that the total order of interior, resp. boundary branching is

$$\sum_{j=1}^{p} \lambda_j = \lambda, \; \sum_{k=1}^{q} \nu_k = \nu.$$

More precisely, to each $\gamma \in \mathcal{A}, x \in D_\nu$ associate a harmonic surface $X = h(\gamma \circ x)$ with $X|_{\partial B} = \gamma \circ x$. Consider the holomorphic vector function of $w = u + iv$

$$(6.2) \qquad F(w) \equiv X_u - iX_v = \partial X.$$

Taking the square of F

$$(6.3) \qquad F^2 = \left(|X_u|^2 - |X_v|^2 \right) - 2i\, X_u \cdot X_v,$$

we obtain another holomorphic function over B.

Now let

$$\eta_\nu^\lambda = \left\{ (\gamma, x) \in \mathcal{A} \times D_\nu \middle| \; F = \prod_{j=1}^{p} (w - w_j)^{\lambda_j} \prod_{k=1}^{q} \left(w - e^{i\phi_k} \right)^{\nu_k} \hat{F}, \right.$$

$$\left. \hat{F} \neq 0 \text{ in } \overline{B}, \; w_j \in B, \; \lambda_j \in I\!N, \; \sum_{j=1}^{p} \lambda_j = \lambda \right\}$$

where the holomorphic function F is associated to the pair (γ, x) by the preceding formulas.

The topologies on $\mathcal{A}, D_\nu, \eta_\nu^\lambda$ are certain Hilbert space topologies defined with reference to $\gamma, \Theta, \{\phi_k\}, \{w_j\}$, and $\hat{F}$ resp.

Define the bundle projection

$$\Pi : \eta \to \mathcal{A}$$

by letting $\Pi_\nu^\lambda : \eta_\nu^\lambda \to \mathcal{A}$ be the map

$$\Pi_\nu^\lambda(\gamma, x) = \gamma.$$

Also introduce the conformality operators K_ν^λ from η_ν^λ into the space

$$Y_\nu^\lambda = \left\{ f \mid f = \prod_{j=1}^{p} (w - w_j)^{2\lambda_j} \prod_{k=1}^{q} (w - e^{i\phi_k})^{2\nu_k} \hat{F}^2, \hat{F} \text{ holomorphic} \right\}$$

of holomorphic functions having (at least) 2λ interior zeros w_j with multiplicities $2\lambda_j$ and 2ν zeroes $e^{i\phi_k} \in \partial B$ with multiplicities $2\nu_k$, by letting

$$K_\nu^\lambda(\gamma, x) = F^2$$

be given by (6.3).

The set of minimal surfaces $\mathcal{M}$ in η now is stratified

$$\mathcal{M} = \bigcup_{\lambda,\nu \in \mathbf{N}_0} \mathcal{M}_\nu^\lambda$$

into subsets

$$\mathcal{M}_\nu^\lambda = \left(K_\nu^\lambda\right)^{-1}\left(F^2 = 0\right) \subset \eta_\nu^\lambda$$

of minimal surfaces of a prescribed branching type.

In the appropriate Hilbert space topologies, the operator $K_\nu^\lambda : \eta_\nu^\lambda \to Y_\nu^\lambda$ is differentiable. Moreover, its differential at a minimal surface $(\gamma, x) \in \eta_\nu^\lambda$ is surjective.

By the implicit function theorem this implies the first part of the following "Index Theorem" of Böhme and Tromba. (Cp. Böhme and Tromba [1, Theorem 3.39]. The case $\nu \neq 0$ was completed by U. Thiel [2].)

__Theorem 6.11:__ $\mathcal{M}_\nu^\lambda$ is a differentiable submanifold of η_ν^λ. The restriction

$$\Pi_\nu^\lambda\big|_{\mathcal{M}_\nu^\lambda} : \mathcal{M}_\nu^\lambda \to \mathcal{A}$$

is Fredholm (i.e. is differentiable and at any $(\gamma, x) \in \mathcal{M}_\nu^\lambda$, $d\left(\Pi_\nu^\lambda\big|_{\mathcal{M}_\nu^\lambda}\right)$ has a closed range R of finite codimension and a finite dimensional kernel N) of Fredholm-index

$$\dim(N) - \operatorname{codim}(R) = 2(2-n)\lambda + (2-n)\nu + 2p + q + 3.$$

__Remark 6.12:__   i) If we normalize with respect to the conformal group on B the Fredholm index of the projection map becomes

$$2(2-n)\lambda + (2-n)\nu + 2p + q.$$

ii)   Note that by the Plateau boundary conditon boundary branch points have even multiplicities $\nu_k \geq 2$. Hence with such a normalization for $n \geq 3$ the index of $\Pi_\nu^\lambda\big|_{\mathcal{M}_\nu^\lambda}$ at a minimal surface X associated with $(\gamma, x) \in \mathcal{M}_\nu^\lambda$ is non-positive and it may equal zero only in case

$$\lambda = \nu = 0, \quad if \ \ n \geq 4$$
$$\nu = 0, \ \lambda_j \leq 1, \quad if \ \ n = 3.$$

iii)   By the Sard-Smale Theorem almost all $\gamma \in \mathcal{A}$ are regular values for all the (countably many) projections $\Pi_\nu^\lambda\big|_{\mathcal{M}_\nu^\lambda}$. Denote this set of regular values $\gamma \in \mathcal{A}$ by $\hat{\mathcal{A}}$. By ii) necessarily the index of $\Pi_\nu^\lambda\big|_{\mathcal{M}_\nu^\lambda}$ at $\gamma \in \mathcal{A}$ equals 0, and $d\left(\Pi_\nu^\lambda\big|_{\mathcal{M}_\nu^\lambda}\right)$ also is injective at γ, i.e. each $\mathcal{M}_\nu^\lambda$ is transversal to the fibre over $\gamma \in \hat{\mathcal{A}}$. In particular, minimal surfaces over $\gamma \in \hat{\mathcal{A}}$ are isolated and stable under perturbations of γ in any manifold $\mathcal{M}_\nu^\lambda$ (with λ, ν fixed). Moreover, by the a-priori estimates of Theorem I.5.1 the set of normalized minimal surfaces spanning a wire γ is compact. However, different sheets $\mathcal{M}_{\nu m}^{\lambda m}$ cannot accumulate at a regular minimal surface X (X would have to have infinite order of branching).

Our Remark 6.12 translates into the following *generic finitenes and stability* result of Böhme-Tromba [1, Theorem 4.14]:

Theorem 6.13:   There exists a dense open set of curves $\tilde{\mathcal{A}} \subset \mathcal{A}$ such that any $\gamma \in \tilde{\mathcal{A}}$ bounds only finitely many minimal surfaces. These surfaces are stable under perturbations of γ. Moreover, they are immersed over $\overline{B}$ if $n \geq 4$, resp. may possess at most simple interior branch points, if $n = 3$.

Theorem 6.13 also provides a (partial) answer to the question whether for a curve Γ the non-degeneracy condition in our Theorem 4.6 is satisfied.

Corollary 6.14:   Let $n \geq 4$, and let $\tilde{\mathcal{A}}$ be as in Theorem 6.13. Then any $\gamma \in \tilde{\mathcal{A}}$ spans only minimal surfaces which are non-degenerate in the sense of Definition 3.2.

Proof: If $n \geq 4$, $\gamma \in \tilde{\mathcal{A}}$ bounds only immersed minimal surfaces $X = h(\gamma \circ x)$ of branching type $\lambda = \nu = 0$, and such that $d\left(\Pi_o^o|_{\mathcal{M}_o^o}\right)$ is an isomorphism of $T_{\gamma,x}\mathcal{M}_o^o = \ker(dK_o^o)$ onto $T_\gamma \mathcal{A}$.

In particular, the restriction $d_x K_o^o$ of dK_o^o to the tangent space $T_x \mathcal{D}_o \cong H^\dagger$ at x in the fibre of normalized surfaces over γ is injective:

$$(6.4) \qquad d_x K_o^o(\gamma, x) \cdot \xi \neq 0 \quad \text{for all } \xi \in T_x \mathcal{D}_o, \ \xi \neq 0.$$

We rewrite this expression in a more convenient way: Note that if X is harmonic also the function

$$\Phi = w^2 \partial X^2 = (u + iv)^2 \cdot (X_u - iX_v)^2 = \left(r^2|X_r|^2 - |X_\phi|^2\right) - 2ir X_r \cdot X_\phi$$

is holomorphic over B. Moreover,

$$K_o^o(\gamma, x) = \partial X^2,$$

whence (6.4) implies that also the holomorphic functions

$$Y = d_x \Phi(\gamma, x) \cdot \xi \neq 0 \quad \text{for all } \xi \in T_x \mathcal{D}_o, \ \xi \neq 0$$

(where d_x again denotes the derivative with respect to x).

In fact, for all such Y the real and imaginary parts separately cannot vanish identically. Else by the Cauchy-Riemann equations $Y \equiv \text{const} = Y(0) = 0$, because Y contains the factor w^2.

Finally, by harmonicity of $im(Y)$ we conclude that

$$im\left(d_x \Phi(\gamma, x) \cdot \xi\right) = 2 d_x \left(X(x)_r \cdot X(x)_\phi\right) \cdot \xi \neq 0 \quad \text{on } \partial B$$

for all $\xi \in T_x \mathcal{D}_o$, $\xi \neq 0$, and we can find $\varsigma \in T_x \mathcal{D}_o$ such that

$$\int\limits_{\partial B} (d_x (X(x)_r \cdot X(x)_\phi) \cdot \xi) \cdot \varsigma \, d\phi =$$

$$= \int\limits_{\partial B} \partial_n X \cdot \frac{d^2}{d\phi^2}\gamma(x)\frac{d}{d\phi}x \cdot \xi \cdot \varsigma d\phi + \int\limits_{\partial B} \partial_n \left(h\left(\frac{d}{d\phi}\gamma(x) \cdot \xi\right)\right) \cdot \frac{d}{d\phi}\gamma(x) \cdot \frac{d}{d\phi}x \cdot \varsigma d\phi$$

$$= d^2 E(x) \left(\xi, \frac{d}{d\phi}x \cdot \varsigma\right) \neq 0.$$

I.e. X is non-degenerate, and the proof is complete.

$$\square$$

The representation of minimal surfaces by smooth submanifolds $\mathcal{M}_\nu^\lambda$ in the bundle η also allowed Tromba to apply a variant of degree theory to the Plateau problem. In this way, in 1982, he was able to give the first proof of the "last" Morse inequality using a geometric notion of "index" for minimal surfaces spanning a generic curve Γ in $I\!R^n$, $n \geq 3$, cf. Tromba [2]:

__Theorem 6.15 :__   For $n \geq 3$ let $\tilde{\mathcal{A}}$ be as in the statement of Theorem 6.13. Then for any $\Gamma \subset \tilde{\mathcal{A}}$ the Morse relation holds:

$$\sum_X (-1)^{i(X)} = 1,$$

where the sum extends over all the (finitely many) minimal surfaces X spanning Γ, and $i(X)$ equals the Morse index of X, in case X is non-degenerate, as defined in Section 4.

However, the "intermediate" Morse inequalities cannot be obtained from degree arguments, and a different approach as outlined in Section 4 was needed to complete the analysis.

B. Surfaces of prescribed constant

mean curvature

III. The existence of surfaces of prescribed constant mean curvature spanning a Jordan curve in $I\!R^3$.

1. The variational problem. Let Γ be a Jordan curve in $I\!R^3$. In part A we studied minimal surfaces spanned by Γ, and we observed that any solution X to the parametric Plateau problem (1.1.1) - (1.1.3) parametrizes a surface of vanishing mean curvature (away from branch points where $\nabla X(w) = 0$).

A natural generalization of the classical Plateau problem therefore is the following question: Given $\Gamma \subset I\!R^3$, $H \in I\!R$, is there a surface X with mean curvature H (for short "H-surface") spanning Γ?

We restrict ourselves to surfaces of the type of the disc B. Introducing isothermal coordinates over B on such a surface X we derive the *parametric form of the Plateau problem for surfaces of constant mean curvature:*

(1.1) $$\triangle X = 2H X_u \wedge X_v \quad \text{in} \quad B$$
(1.2) $$|X_u|^2 - |X_v|^2 = 0 = X_u \cdot X_v \quad \text{in} \quad B$$
(1.3) $$X|_{\partial B} : \partial B \to \Gamma \text{ is a parametrization of } \Gamma.$$

Here $a \wedge b = (a^2 b^3 - b^2 a^3, \, a^3 b^1 - b^3 a^1, \, a^1 b^2 - b^1 a^2)$ denotes the exterior product of $a = (a^1, a^2, a^3)$, $b = (b^1, b^2, b^3) \in I\!R^3$. Formally, these are the Euler-Lagrange equations corresponding to the functional

(1.4) $$D_H(X) = D(X) + 2H V(X),$$

where

(1.5) $$V(X) = 1/3 \int_B X_u \wedge X_v \cdot X \, dw$$

denotes the "volume" of X. In fact, $V(X)$ measure the algebraic volume enclosed in the cone segment consisting of all lines joining points $X(w)$ on X with the origin. This is immediate for surfaces whose coordinates are linear functions: $X(u,v) = au + bv + c$, $a, b, c \in I\!R^3$. For smooth surfaces this geometric interpretation of the volume may be obtained by approximation of X with polyhedral surfaces.

Likewise, we may also regard $V(X)$ as the algebraic volume bounded by X and the surface X_o consisting of all line segments joining points on $\Gamma = X(\partial B)$ with the origin.

However, we need not be too concerned about the geometric meaning of V, and simply take formula (1.5) as a *definition*.

Remark 1.1:   i) $V(X)$ is well-defined and tri-linear, in particular analytic on the space $H^{1,2} \cap L^{\infty}(B; \mathbb{R}^3)$.

ii)   For any $X, \varphi, \psi \in H^{1,2} \cap L^{\infty}(B; \mathbb{R}^3)$ there holds the expansion

(1.6) $$V(X + \varphi) = V(X) + \langle dV(X), \varphi \rangle + 1/2\, d^2 V(X)(\varphi, \varphi) + V(\varphi),$$

where

(1.7) $$\langle dV(X), \varphi \rangle = 1/3 \left[\iint_B (\varphi_u \wedge X_v + X_u \wedge \varphi_v) \cdot X\, dw + \int_B X_u \wedge X_v \cdot \varphi dw \right],$$

(1.8)
$$d^2 V(X)(\varphi, \psi) = 1/3 \left[\iint_B (\varphi_u \wedge \psi_v + \psi_u \wedge \varphi_v) \cdot X\, dw \right.$$
$$\left. + \int_B (\varphi_u \wedge X_v + X_u \wedge \varphi_v) \cdot \psi + (\psi_u \wedge X_v + X_u \wedge \psi_v) \cdot \varphi dw \right].$$

iii)   If $\rho, \varphi, \psi \in C^2(\overline{B}; \mathbb{R}^3)$ and at least one function $\rho, \varphi,$ or ψ vanishes on ∂B, then by partial integration and antisymmerty $a \wedge b = -b \wedge a$:

$$\int_B (\varphi_u \wedge \psi_v + \psi_u \wedge \varphi_v) \cdot \rho dw =$$

(1.9)
$$= -\int_B \varphi \wedge \psi_v \cdot \rho_u + \psi_u \wedge \varphi \cdot \rho_v\, dw + \int_{\partial B} (\varphi \wedge u\psi_v + v\psi_u \wedge \varphi) \cdot \rho do$$
$$= \int_B (\rho_u \wedge \psi_v + \psi_u \wedge \rho_v) \cdot \varphi dw + \int_{\partial B} \varphi \wedge \frac{\partial}{\partial \phi}\psi \cdot \rho\, do$$
$$= \int_B (\rho_u \wedge \psi_v + \psi_u \wedge \rho_v) \cdot \varphi\, dw = \int_B (\varphi_u \wedge \rho_v + \rho_u \wedge \varphi_v) \cdot \psi\, dw.$$

Here, $\frac{\partial}{\partial \phi}\psi = u\psi_v - v\psi_u$ as usual denotes the derivative of ψ in "angular", i.e. counter-clockwise tangent direction along ∂B. Hence, if either φ, ψ or $\rho \equiv 0$ on ∂B, the boundary integral vanishes.

In particular, if $X \in C^2(\overline{B}; \mathbb{R}^3)$, $\varphi, \psi \in C_o^{\infty}(B; \mathbb{R}^3)$ there holds

(1.10') $$\langle dV(X), \varphi \rangle = \int_B X_u \wedge X_v \cdot \varphi dw,$$

(1.10") $$d^2 V(X)(\varphi, \psi) = \int_B (\varphi_u \wedge \psi_v + \psi_u \wedge \varphi_v) \cdot X dw.$$

iv) V is invariant under orientation-preserving reparametrizations of X :
Let $X \in H^{1,2} \cap L^\infty(B; I\!R^3)$, and let $g \in C^1(\overline{B}; I\!R^2)$ be a diffeomorphism of
B onto a domain $\hat{B}$ with $\det(dg) = g_u^1 g_v^2 - g_u^2 g_v^1 > 0$, $\hat{X} = X \circ g^{-1} \in$
$H^{1,2} \cap L^\infty(\hat{B}, I\!R^3)$. Then

(1.11)
$$V(\hat{X}) = 1/3 \int_{\hat{B}} \hat{X}_u \wedge \hat{X}_v \cdot \hat{X} dw$$

$$= 1/3 \int_B X_u \wedge X_v \cdot X \det \left(d(g^{-1}) \circ g \right) | \det(dg) | dw = V(X).$$

v)   If $X \in \mathcal{C}(\Gamma) \cap C^2(B; I\!R^3)$ is a stationary point of D_H with respect to
variations of the dependent and independent variables, cp. Lemma I.2.2, from (1.10)
and (1.11) we obtain the weak form of (1.1)

(1.12)
$$\langle dD_H(X), \varphi \rangle = \int_B \nabla X \nabla \varphi + 2H\, X_u \wedge X_v \cdot \varphi dw$$

$$= \int_B [-\triangle X + 2H\, X_u \wedge X_v] \cdot \varphi dw = 0, \quad \forall\, \varphi \in C_o^\infty,$$

resp. the conformality relations, cp. Lemma I.2.4:

(1.13)
$$\frac{d}{d\epsilon} D_H \left(X \circ (id + \epsilon\tau)^{-1} \right) \Big|_{\epsilon=0}$$

$$= \frac{d}{d\epsilon} D \left(X \circ (id + \epsilon\tau)^{-1} \right) \Big|_{\epsilon=0} = 0, \ \forall\, \tau \in C^1(\overline{B}; I\!R^2).$$

I.e., X is an $H-$surface in conformal representation.

Remark 1.1. v) justifies our claim that the parametric $H-$surface problem (1.1)-
(1.3) formally corresponds to the Euler-Lagrange equations of D_H on $\mathcal{C}(\Gamma)$.

To make this precise we now analyze the volume functional V in detail.

2. The volume functional.   The basic tool in this section is the following isoperimetric inequality for closed surfaces in $I\!R^3$, cf. Radó [4].

Theorem 2.1:   Let $X, Y \in H^{1,2} \cap L^\infty(B; I\!R^3)$ satisfy $X - Y \in H_o^{1,2}(B; I\!R^3)$. Then

$$36\pi \, |V(X) - V(Y)|^2 \le [D(X) + D(Y)]^3,$$

and the constant 36π is best possible.

Remark 2.2:   i)   The constant 36π is achieved for example if $X = X_+$, $Y = X_-$ where

$$X_\pm(u, v) = \frac{1}{1 + u^2 + v^2} \left(2u, 2v, \pm \left(1 - u^2 - v^2 \right) \right)$$

denote stereographic representations of an upper and a lower hemi-sphere of radius 1 centered at 0.

ii)   Recall that V is invariant under orientation-preserving changes of parameters. Moreover, by the ϵ−conformality Theorem I.2.1 of Morrey we may introduce coordinates on X to achieve $D(X) \le (1 + \epsilon)A(X)$ for any given $\epsilon > 0$. Hence Theorem 2.1 implies the estimate

$$36\pi |V(X) - V(Y)|^2 \le [A(X) + A(Y)]^3$$

for all $X, Y \in H^{1,2} \cap L^\infty(B; I\!R^3)$ with the property that there exists an oriented diffeomorphism g of $\overline{B}$ onto itself such that $X|_{\partial B} = Y \circ g|_{\partial B}$.

Theorem 2.1 and Remark 1.1 have important consequences. The following result (like many results on the analytic properties of H−surfaces) is due to H.C. Wente [1] .

Theorem 2.3:   i)   For any $X \in H^{1,2} \cap L^\infty(B; I\!R^3)$ V continously extends to an analytic functional on $X + H_o^{1,2}(B; I\!R^3)$.

V has the expansion in direction $\varphi \in H_o^{1,2}(B; I\!R^3)$:

(2.1) $V(X + \varphi) = V(X) + \langle dV(X), \varphi \rangle + (1/2 \,) d^2 V(X)(\varphi, \varphi) + V(\varphi).$

ii)   The first variation dV given by

(2.2) $\langle dV(X), \varphi \rangle = \displaystyle\int_B X_u \wedge X_v \cdot \varphi dw, \; \forall \varphi \in H_o^{1,2} \cap L^\infty(B; I\!R^3)$

continuously extends to a map $dV : H^{1,2}(B; I\!R^3) \to \left(H_o^{1,2}(B; I\!R^3) \right)^*$ which satisfies the estimate

(2.3) $|\langle dV(X), \varphi \rangle| \le cD(X) \, D(\varphi)^{1/2} ,$

and is weakly continuous in the sense that

(2.4)
$$X_m \xrightarrow{w} X \text{ in } H^{1,2}(B; I\!R^n) \Rightarrow \langle V(X_m), \varphi \rangle \to \langle dV(X), \varphi \rangle,$$
$$\forall \, \varphi \in H_o^{1,2}(B; I\!R^3).$$

iii)   The second variation d^2V given by

(2.5)
$$d^2V(X)(\varphi, \psi) = \int_B (\varphi_u \wedge \psi_v + \psi_u \wedge \varphi_v) \cdot X \, dw, \quad \forall \, \varphi, \psi \in H_o^{1,2}(B; I\!R^3)$$

continuously extends to a map $d^2V : H^{1,2}(B; I\!R^3) \to \left(H_o^{1,2} \times H_o^{1,2}(B; I\!R^3) \right)^*$ which satisfies the estimate

(2.6)
$$|d^2V(X)(\varphi, \psi)| \le c(D(X) D(\varphi) D(\psi))^{1/2}$$

and is weakly continuous in the sense that

(2.7)
$$X_m \xrightarrow{w} X \text{ in } H^{1,2}(B; I\!R^3) \Rightarrow d^2V(X_m)(\varphi, \psi) \to d^2V(X)(\varphi, \psi),$$
$$\forall \, \varphi, \psi \in H_o^{1,2}(B; I\!R^3).$$

Moreover, $d^2V(X)$ for fixed $X \in H^{1,2}(B; I\!R^3)$ is a completely continuous bilinear form on $H_o^{1,2}(B; I\!R^3)$ in the sense that

(2.8)
$$\varphi_m \xrightarrow{w} \varphi, \ \psi_m \xrightarrow{w} \psi \text{ in } H_o^{1,2}(B; I\!R^3)$$
$$\Longrightarrow d^2V(X)(\varphi_m, \psi_m) \to d^2V(X)(\varphi, \psi).$$

iv)   If $X_m, X \in \mathcal{C}(\Gamma)$ and $X_m \xrightarrow{w} X$ in $H^{1,2}(B; I\!R^3)$ while $X_m \to X$ uniformly on $\overline{B}$, and if $\varphi_m \xrightarrow{w} \varphi, \ \psi_m \xrightarrow{w} \psi$ in $H_o^{1,2}(B; I\!R^3)$, then

(2.9)
$$V(X_m) \to V(X) \,,$$

(2.10)
$$\langle dV(X_m), \varphi_m \rangle \to \langle dV(X), \varphi \rangle \,,$$

(2.11)
$$d^2V(X_m)(\varphi_m, \psi_m) \to d^2V(X)(\varphi, \psi)$$

as $m \to \infty$.

Proof: By (1.6), (1.10) formulas (2.1), (2.2), (2.5) hold for $X \in C^2(\overline{B}; I\!R^3)$, $\varphi, \psi \in C_o^\infty(B; I\!R^3)$. By uniform continuity of the integrals $\int_B X_u \wedge X_v \cdot \varphi \, dw$ with respect to $X \in H^{1,2}(B; I\!R^3)$, $\varphi \in H^{1,2} \cap L^\infty(B; I\!R^3)$ it is also clear that dV continuously extends to $dV : H^{1,2}(B; I\!R^3) \to (H_o^{1,2} \cap L^\infty(B; I\!R^3))^*$. Similarly, by (1.9) and (2.5) d^2V extends to a map

$$d^2V : H^{1,2}(B; I\!R^3) \to \left(\left[H^{1,2} \cap L^\infty(B; I\!R^3) \right]^2 \right)^* .$$

Once we have established (2.3), (2.6) , moreover, $dV(X)$ extends to a continuous linear functional on $H_o^{1,2}(B; I\!R^3)$ while $d^2V(X)$ continuously extends to a bilinear form on $\left[H_o^{1,2}(B; I\!R^3) \right]^2$ as claimed.

(2.3) and (2.6) are deduced from Theorem 2.1 as follows:

For $\quad X = (X^1, X^2, X^3) \in H^{1,2} \cap L^\infty(B; I\!\!R^3) \quad$ and $\quad \varphi = (\varphi^1, \varphi^2, \varphi^3) \in H_o^{1,2} \cap L^\infty(B; I\!\!R^3)$ let

$$Y = \left(\frac{X^1}{D(X)^{1/2}}, \frac{X^2}{D(X)^{1/2}}, 0 \right), Z = \left(\frac{X^1}{D(X)^{1/2}}, \frac{X^2}{D(X)^{1/2}}, \frac{\varphi^3}{D(\varphi)^{1/2}} \right).$$

Note that $\quad V(Y) = 0, \; D(Y) \leq 1, \; D(Z) \leq 2.$ Applying the isoperimetric inequality Theorem 2.1 to $\quad Y \quad$ and $\quad Z \quad$ we obtain

$$|V(Z)|^2 \leq \frac{3}{4\pi}.$$

By antisymmetry of the volume element $\quad a \wedge b \cdot c \quad$ now $\quad V \quad$ is also trilinear in the components of $\quad Z = (Z^1, Z^2, Z^3).$ Multiplying by $\quad D(X)^2 \, D(\varphi) \quad$ we hence find that

$$|V(X^1, X^2, \varphi^3)|^2 = |\langle dV(X), (0, 0, \varphi^3) \rangle|^2 \leq \frac{3}{4\pi} D(X)^2 D(\varphi).$$

Repeating the argument for the remaining two components of $\quad \varphi \quad$ (2.3) follows.

To see (2.6) let $\quad X, \varphi$ as above , $\quad \psi \in H_o^{1,2} \cap L^\infty(B; I\!\!R^3),$ and set

$$Y = \left(\frac{X^1}{D(X)^{1/2}}, 0, 0 \right), \quad Z = \left(\frac{X^1}{D(X)^{1/2}}, \frac{\varphi^2}{D(\varphi)^{1/2}}, \frac{\psi^3}{D(\psi)^{1/2}} \right).$$

Then the above reasoning gives (denoting e.g. $\quad (0, \varphi^2, 0) = \varphi^2$ for brevity)

$$|d^2 V(X)(\varphi^2, \psi^3)|^2 = |V(Z)|^2 D(X) D(\varphi) D(\psi)$$

$$\leq \frac{16}{9\pi} D(X) D(\varphi) D(\psi),$$

and (2.6) follows by trilinearity of $\quad V$:

$$|d^2 V(X)(\varphi, \psi)| \leq \sum_{j \neq k} |d^2 V(X)(\varphi^j, \psi^k)|$$

$$\leq c (D(X) D(\varphi) D(\psi))^{1/2}.$$

Finally, also $\quad V \quad$ extends to a trilinear functional in the components of $\quad \varphi = (\varphi^1, \varphi^2, \varphi^3) \in H_o^{1,2}(B; I\!\!R^3)$ by the estimate

$$|V(\varphi)|^2 = |V(\varphi^1, \varphi^2, \varphi^3)|^2 \leq \frac{1}{36\pi} D(\varphi)^3,$$

and (2.1) follows.

Let us now establish the continuity properties asserted in (2.4), (2.7), (2.8).

Proof of (2.4): By (1.9) for any $\quad \varphi \in C_o^\infty(B; I\!\!R^3)$

$$\langle dV(X_m), \varphi \rangle = \int_B X_{mu} \wedge X_{mv} \cdot \varphi \, dw$$

$$= 1/2 \int_B (\varphi_u \wedge X_{mv} + X_{mu} \wedge \varphi_v) \cdot X_m \, dw.$$

Note that by compactness of the embedding $H^{1,2}(B; I\!\!R^3) \hookrightarrow L^2(B; I\!\!R^3)$ $X_m \to$ X strongly in L^2 while the products $\varphi_u \wedge X_{mv}$ etc. remain uniformly bounded in L^2. Hence with error $o(1) \to o(m \to \infty)$

$$\langle dV(X_m), \varphi \rangle = 1/2 \int\limits_B (\varphi_u \wedge X_{mv} + X_{mu} \wedge \varphi_v) \cdot X \, dw + o(1)$$

and by weak convergence $X_m \overset{w}{\to} X$ in $H^{1,2}$ the right hand side tends to

$$1/2 \int\limits_B (\varphi_u \wedge X_v \wedge X_u \wedge \varphi_v) \cdot X dw = \langle dV(X), \varphi \rangle$$

as required.

This proves (2.4) for smooth φ. The general case follows by density of $C_o^\infty(B; I\!\!R^3)$ in $H_o^{1,2}(B; I\!\!R^3)$ and the uniform boundedness of the family $\{dV(X_m)\}$ of linear functionals on $H_o^{1,2}(B; I\!\!R^3)$ which results as a consequence of (2.3) and the boundedness of weakly convergent sequences in $H^{1,2}(B; I\!\!R^3)$.

Proof of (2.7), (2.8):   Similarly, for $\varphi, \psi \in C_o^\infty(B; I\!\!R^3)$

$$d^2V(X_m)(\varphi, \psi) = \int\limits_B (\varphi_u \wedge \psi_v + \psi_u \wedge \varphi_v) \cdot X_m \, dw$$

$$\to \int\limits_B (\varphi_u \wedge \psi_v + \psi_u \wedge \phi_v) \cdot X \, dw$$

$$= dV(X)(\varphi, \psi),$$

and the general case follows using (2.6). Likewise, for $X \in C^1(\overline{B}; I\!\!R^3)$, $\phi_m \overset{w}{\to} \varphi$, $\psi_m \overset{w}{\to} \psi$ in $H_o^{1,2}(B; I\!\!R^3)$ by arguments as in the proof of (2.4)

$$d^2V(X)(\varphi_m, \psi_m) = \int\limits_B (\varphi_{mu} \wedge X_v + X_u \wedge \varphi_{mv}) \cdot \psi_m dw$$

$$\to \int\limits_B (\varphi_u \wedge X_v + X_u \wedge \varphi_v) \cdot \psi dw$$

$$= d^2V(X)(\varphi, \psi),$$

and the general case follows by density of $C^1(\overline{B}; I\!\!R^3)$ in $H^{1,2}(B; I\!\!R^3)$ and (2.6).

The **proof of iv)** is slightly more delicate. Let $X_m, X \in \mathcal{C}(\Gamma)$ and assume that $X_m \overset{w}{\to} X$ weakly on $H^{1,2}(B; I\!\!R^3)$ and uniformly on $\overline{B}$.

Then upon integrating by parts we find

$$6[V(X_m) - V(X)] = 2 \int_B X_{mu} \wedge X_{mv} \cdot X_m - X_u \wedge X_v \cdot X \, dw$$

$$= \int_B ((X_m - X)_u \wedge X_{mv} + X_{mu} \wedge (X_m - X)_v) \cdot X_m \, dw$$

$$+ \int_B ((X_m - X)_u \wedge X_v + X_u \wedge (X_m - X)_v) \cdot X_m dw$$

$$+ 2 \int_B X_u \wedge X_v \cdot (X_m - X) \, dw$$

$$= \int_B (X_{mu} \wedge (X_m + X)_v + (X_m + X)_u \wedge X_{mv} + 2 X_u \wedge X_v) \cdot (X_m - X) \, dw$$

$$+ \int_{\partial B} (X_m - X) \wedge [(u \cdot X_{mv} - v \cdot X_{mu}) + (u \cdot X_v - v \cdot X_u)] \cdot X_m \, do$$

$$\leq c \cdot (D(X_m) + D(X)) \cdot \|X_m - X\|_{L^\infty(B)} +$$

$$+ c \cdot \|X_m - X\|_{L^\infty(\partial B)} \int_{\partial B} \left| \frac{\partial}{\partial \phi} X_m \right| + \left| \frac{\partial}{\partial \phi} X \right| do$$

$$\to 0,$$

since $\quad \|X_m\|_{L^\infty(\partial B)} \leq c(\Gamma), \quad \int_{\partial B} \left| \frac{\partial}{\partial \phi} X_m \right| do \ = \int_{\partial B} \left| \frac{\partial}{\partial \phi} X \right| do = L(\Gamma) < \infty.$ This proves (2.9).

(2.10) follows since by (1.9) if $\quad \varphi_m \xrightarrow{w} \varphi \in H_o^{1,2}(B; \mathbb{R}^3) :$

$$2[\langle dV(X_m), \varphi_m \rangle - \langle dV(X), \varphi \rangle] = 2 \int_B (X_{mu} \wedge X_{mv} \cdot \varphi_m - X_u \wedge X_v \cdot \varphi) \, dw$$

$$= \int_B ((X_m - X)_u \wedge (X_m + X)_v + (X_m + X)_u \wedge (X_m - X)_v) \cdot \varphi_m \, dw$$

$$+ \int_B 2 X_u \wedge X_v \cdot (\varphi_m - \varphi) dw$$

$$= \int_B (\varphi_{mu} \wedge (X_m + X)_v + (X_m + X)_u \wedge \varphi_{mv}) \cdot (X_m - X) dw$$

$$+ \int_B (X_u \wedge (\varphi_m - \varphi)_v + (\varphi_m - \varphi)_u \wedge X_v) \cdot X dw$$

$$\to \ 0 \quad (m \to \infty).$$

A similar reasoning shows (2.11).

□

Remark 1.1 and Theorem 2.3 imply the following

Remark 2.4:   i) D_H extends to an analytic functional on
$(H^{1,2} \cap L^\infty(B; I\!\!R^3)) + H_o^{1,2}(B; I\!\!R^3)$ with

$$(2.12) \quad D_H(X + \varphi) = D_H(X) + \langle dD_H(X), \varphi \rangle + (1/2) d^2 D_H(X)(\varphi, \varphi)$$
$$+ 2H \, V(\varphi)$$
$$(2.13) \quad\quad\quad = D_H(X) + D_H(\varphi) + \langle dD_H(X), \varphi \rangle + H d^2 V(X)(\varphi, \varphi),$$

for any $X, \varphi \in (H^{1,2} \cap L^\infty(B; I\!\!R^3)) + H_o^{1,2}(B; I\!\!R^3)$.

   ii) X is conformal and weakly solves (1.1) iff

$$dD_H(X) = 0 \in \left(H_o^{1,2}(B; I\!\!R^3)\right)^*,$$
$$\frac{d}{d\epsilon} D_H(X \circ (id + \epsilon \tau)^{-1})|_{\epsilon=0} = \frac{d}{d\epsilon} D(X \circ (id + \epsilon \tau)^{-1}|_{\epsilon=0} = 0, \; \forall \; \tau \in C^1(\overline{B}; I\!\!R^2),$$

i.e. iff X is stationary for D_H on $\mathcal{C}(\Gamma)$ in the sense of Lemma I.2.2.

3. "Small" solutions.   The existence of relative minimizers of D_H in $\mathcal{C}(\Gamma)$ has been obtained under various geometric conditions relating Γ and H and under certain a priori restrictions on admissible solution surfaces.

By Remark 2.4. ii) a relative minimizer $X \in \mathcal{C}(\Gamma)$ of D_H weakly solves (1.1) -(1.3). Later we shall see that X is as regular as the data permit and hence also solves (1.1) - (1.3) classically.

Let us first state the following result of Hildebrandt [2] which improves earlier results of Heinz [1] and Werner [1]:

Theorem 3.1:   Let $\Gamma \subset B_R(0) \subset I\!\!R^3$ be a rectifiable Jordan curve, and let $H \in I\!\!R$ satisfy

$$(3.1) \qquad\qquad |H|R \le 1.$$

Then there exists a solution $\underline{X} \in \mathcal{C}(\Gamma)$ to (1.1) -(1.3) characterized by the conditons

$$(3.2) \qquad ||\underline{X}||_{L^\infty} \le R,$$

$$(3.3) \qquad D_H(\underline{X}) = \min\{D_H | X \in \mathcal{C}(\Gamma),\ ||X||_{L^\infty} \le \frac{1}{|H|} \le \infty\}\ .$$

Moreover, if $|H|R < 1$, $\underline{X}$ is a relative minimum of D_H in $\mathcal{C}(\Gamma)$ with respect to the topology of $H^{1,2} \cap L^\infty(B; I\!\!R^3)$.

Heinz' non-existence result Theorem 4.1 below shows that the condition (3.1) cannot be improved when Γ is a planar circle. For lengthy curves, however, Theorem 3.4 below by Wente and Steffen may give better results.

Let
$$\mathcal{C}_H = \{X \in \mathcal{C}(\Gamma) |\ ||X||_{L^\infty} \le \frac{1}{|H|}\}.$$

Note that $\mathcal{C}_H \ne \emptyset$ for $H \in I\!\!R$ satisfying (3.1).

Lemma 3.2: D_H is coercive on $\mathcal{C}_H$ with respect to the $H^{1,2}(B; I\!\!R^3)$-norm.

Proof: Note that for $X \in \mathcal{C}_H$ there holds

$$(3.4) \qquad \begin{aligned} \left|\frac{2H}{3}X_u \wedge X_v \cdot X\right| &\le \frac{2}{3}|H|\,||X||_{L^\infty}\,|X_u|\,|X_v| \\ &\le \frac{1}{3}|\nabla X|^2, \quad \text{a.e. on } B. \end{aligned}$$

Hence for $X \in \mathcal{C}_H$

$$(3.5) \qquad \frac{1}{3}D(X) \le D_H(X) \le \frac{5}{3}D(X)$$

and coerciveness of D_H follows as in example I.3.5.

$\square$

Lemma 3.3: D_H is lower semi-continuous on $\mathcal{C}_H$ with respect to weak convergence in $H^{1,2}(B;\mathbb{R}^3)$.

Proof: Suppose $X_m \xrightarrow{w} X \in \mathcal{C}_H$ weakly in $H^{1,2}(B;\mathbb{R}^3)$. By the Rellich-Kondrakov-Theorem $X_m \to X$ strongly in $L^2(B;\mathbb{R}^3)$ and a.e. on B. By Egorov's theorem then, for any $\delta > 0$ there is a set $B^\delta \subset B$ of measure $|B^\delta| < \delta$ such that $X_m \to X$ uniformly on $B \backslash B^\delta$.

Hence by (3.4) and example I.3.4. ii)

$$
\begin{aligned}
D_H(X_m) &= 1/2 \int_B |\nabla X_m|^2 dw + (2H/3) \int_B X_{m_u} \wedge X_{m_v} \cdot X_m dw \\
&\geq 1/2 \int_{B \backslash B^\delta} |\nabla X_m|^2 dw + (2H/3) \int_{B \backslash B^\delta} X_{m_u} \wedge X_{m_v} \cdot X_m dw \\
&\geq 1/2 \int_{B \backslash B^\delta} |\nabla X|^2 dw + (2H/3) \int_{B \backslash B^\delta} X_u \wedge X_v \cdot X dw - o(1) \\
&= D_H(X) - \left(1/2 \int_{B^\delta} |\nabla X|^2 dw + (2H/3) \int_{B^\delta} X_u \wedge X_v \cdot X dw \right) - o(1)
\end{aligned}
$$

where $o(1) \to 0$ as $m \to \infty$, for any fixed $\delta > 0$.

By absolute continuity of the Lebesgue integral

$$
1/2 \int_{B^\delta} |\nabla X|^2 dw + (2H/3) \int_{B^\delta} X_u \wedge X_v \cdot X dw \to 0 \quad (\delta \to 0),
$$

and we obtain that

$$
\liminf_{m \to \infty} D_H(X_m) \geq D_H(X)
$$

as desired.

$\square$

Proof of Theorem 3.1:   Also let

$$
\mathcal{C}_H^* = \mathcal{C}_H \cap \mathcal{C}^*(\Gamma).
$$

By conformal invariance of D_H :

$$
\beta_H := \inf_{X \in \mathcal{C}_H^*} D_H(X) = \inf_{X \in \mathcal{C}_H} D_H(X).
$$

Note that by Proposition I.4.5 $\mathcal{C}_H^*$ is weakly closed in $H^{1,2}(B; \mathbb{R}^3)$. Moreover, D_H is coercive and weakly lower semi-continuous on $\mathcal{C}_H^*$ with respect to $H^{1,2}(B; \mathbb{R}^3)$. By Theorem I.3.3 β_H is attained at $\underline{X} \in \mathcal{C}_H^*$. We show that $\underline{X}$ for all $|H|R < 1$ is a relative minimizer of D_H with respect to the topology of $H^{1,2} \cap L^\infty(B; \mathbb{R}^3)$.

Let $\varphi \in H_o^{1,2} \cap L^\infty(B; \mathbb{R}^3)$ satisfy $0 \le \varphi \le 1$. Then for $\epsilon \in [0,1]$

$$X_\epsilon = \underline{X} - \epsilon\varphi\underline{X} \in \mathcal{C}_H,$$

and by minimality of $D_H(\underline{X})$ and (3.4) we obtain

$$0 \le \frac{d}{d\epsilon} D_H(X_\epsilon)|_{\epsilon=0} = - \int_B \nabla\underline{X} \, \nabla(\varphi\underline{X}) + 2H\underline{X}_u \wedge \underline{X}_v \cdot \varphi\underline{X} dw$$

$$= -\int_B \nabla\left(\frac{|\underline{X}|^2}{2}\right) \nabla\varphi + \left[|\nabla\underline{X}|^2 + 2H\underline{X}_u \wedge \underline{X}_v \cdot \underline{X}\right] \varphi \, dw$$

$$\le -\int_B \nabla\left(\frac{|\underline{X}|^2}{2}\right) \nabla\varphi dw.$$

I.e. the function $|\underline{X}|^2 \in H^{1,2} \cap L^\infty(B; \mathbb{R}^3)$ is a weak sub-solution to the Poisson equation

$$-\Delta(|\underline{X}|^2) \le 0.$$

Choosing $\varphi = \left(|\underline{X}|^2 - R^2\right)_+ = \sup\left(|\underline{X}|^2 - R^2, 0\right) \in H_o^{1,2} \cap L^\infty$ we obtain that

$$\int_B \left|\nabla\left(|\underline{X}|^2 - R^2\right)_+\right|^2 dw = 0,$$

i.e. $|\underline{X}| \le R$ a.e. on B, and (3.2) holds. It follows that $\underline{X}$ lies interior to $\mathcal{C}_H$ with respect to the topology of $H^{1,2} \cap L^\infty(B; \mathbb{R}^3)$ and is a relative minimizer of D_H in this class. In particular, by Remark 2.4. ii) $\underline{X}$ solves (1.1)-(1.3).

The case $|H|R = 1$:   Choose a sequence $\{H_m\}$ of numbers $H_m \to H$, $|H_m|R < 1$, and let $\underline{X}_m$ be a corresponding sequence of H_m-surfaces characterized by (3.2), (3.3). In particular, by (3.2) $\underline{X}_m \in \mathcal{C}_H^*$. By Lemmata 3.2, 3.3 therefore we may assume that

$$\underline{X}_m \stackrel{w}{\rightharpoonup} X \in \mathcal{C}_H^* \text{ weakly in } H^{1,2}(B; \mathbb{R}^3)$$

and by (3.3), letting $\underline{X} \in \mathcal{C}_H \subset \mathcal{C}_{H_m}$ be a minimizer of D_H :

$$\beta_H \le D_H(X) \le \liminf_{m \to \infty} D_H(\underline{X}_m) = \liminf_{m \to \infty} D_{H_m}(\underline{X}_m)$$

$$\le \liminf_{m \to \infty} D_{H_m}(\underline{X}) = D_H(\underline{X}) = \beta_H.$$

Hence X minimizes D_H in $\mathcal{C}_H$, and in particular

$$\frac{d}{d\epsilon} D_H(X \circ (id + \epsilon\tau)^{-1})|_{\epsilon=0} = 0, \quad \forall\, \tau \in C^1(\overline{B}; I\!\!R^2),$$

i.e. X is conformal, cp. Remark 2.4 and Lemma I.2.4.

Finally, for any $\varphi \in H_o^{1,2}(B; I\!\!R^3)$

$$0 = \langle dD_{H_m}(\underline{X}_m), \varphi \rangle = \langle dD(\underline{X}_m), \varphi \rangle + 2H_m \langle dV(\underline{X}_m), \varphi \rangle$$
$$\to \langle dD(X), \varphi \rangle + 2H \langle dV(X), \varphi \rangle = \langle dD_H(X), \varphi \rangle$$

as $m \to \infty$, by weak continuity of $dD, dV : H^{1,2}(B; I\!\!R^3) \to \left(H_o^{1,2}(B; I\!\!R^3)\right)^*$ and uniform boundedness $|\langle dV(\underline{X}_m), \varphi \rangle| \le c\, D(\underline{X}_m)D(\varphi)^{1/2} \le c$. I.e. $dD_H(X) = 0 \in \left(H_o^{1,2}(B; I\!\!R^3)\right)^*$, and X weakly solves (1.1)-(1.3).

$\square$

The next existence result is due to Steffen [3] , improving an earlier result of Wente [1] .

Theorem 3.4: Let Γ be a Jordan curve in $I\!\!R^3, H \in I\!\!R$. Suppose that for some $X_o \in \mathcal{C}(\Gamma)$ there holds

(3.6) $$H^2 D(X_o) < 2\pi/3.$$

Then there exists a solution $\underline{X}$ to (1.1) -(1.3) which is a relative minimizer of D_H in the class $\{X \in \mathcal{C}(\Gamma)\,|\,D(X) \le 5D(X_o)\}$.

The proof of Theorem 3.4 will be given in a later section, cp. Theorem IV.4.13. Note that the constant $2/3$ in (3.6) is not optimal. In fact, Remark IV.4.14 shows that the assertion of the theorem remains true provided $H^2 D(X_o) < 2\pi/3 + \epsilon$ for some $\epsilon > 0$ depending on Γ. It is conjectured that it suffices to assume that $H^2 D(X_o) < \pi$, which is optimal in the case of a circle.

4. Heinz' non-existence result. The necessity of the smallness condition in the existence theorems of the preceding section is illustrated by the following result of Heinz [2]:

Theorem 4.1: Let $\Gamma \subset \mathbb{R}^3$ be a rectificiable Jordan curve of length $L(\Gamma)$, and suppose that for some $X_o \in \mathcal{C}(\Gamma)$ and some unit vector $n_o \in \mathbb{R}^3$ the number

$$c_o = \int_B X_{o_u} \wedge X_{o_v} \cdot n_o dw > 0.$$

Then if $H \in \mathbb{R}$ satisfies $|H| > (L(\Gamma)/2c_o)$ there is no solution to (1.1)-(1.3).

Remark 4.2: In particular , if $\Gamma = \partial B$ is the unit circle in $\mathbb{R}^2$ we may let $X_o(w) \equiv w$, $n_o = (0, 0, 1)$. Then $L(\Gamma) = 2\pi$, $c_o = \pi$, and it follows that Γ cannot span $H-$surfaces with curvature $|H| > 1$.

Proof: We present the proof in case $\Gamma \in C^{1,\alpha}$. Then any $H-$surfaces $X \in \mathcal{C}(\Gamma)$ is of class $C^1(\overline{B}; \mathbb{R}^3)$, cp. Theorem 5.5, and from (1.1)-(1.3) we obtain

$$2H \int_B X_u \wedge X_v \cdot n_o dw = \int_B \triangle X \cdot n_o dw = \int_{\partial B} \partial_n X \cdot n_o do$$

$$\leq \int_{\partial B} |\partial_n X| do = \int_{\partial B} \left| \frac{\partial}{\partial \phi} X \right| do = L(\Gamma).$$

By invariance of the integral

$$c_o = n_o \cdot \int_B X_{o_u} \wedge X_{o_v} dw$$

under orientation preserving changes of parametrization we may assume that $X - X_o \in H_o^{1,2}(B; \mathbb{R}^3)$. But then by (1.9)

$$2 \int_B \left(X_u \wedge X_v - X_{o_u} \wedge X_{o_v} \right) \cdot n_o dw =$$

$$= \int_B [(X - X_o)_u \wedge (X + X_o)_v + (X + X_o)_u \wedge (X - X_o)_v] \cdot n_o dw = 0$$

and it follows that $2c_o|H| \leq L(\Gamma)$.   □

Remark: It is not known whether a similar result holds for the Dirichlet problem (IV.1.1)-(IV.1.2) below.

5. Regularity.   Regularity of weak solutions to (1.1) - (1.3) with minor modifications can be obtained as in the case of minimal surfaces $(H = 0)$ once the following result due to the Wente [5, Lemma 3.1] has been established.

Theorem 5.1:   Let $\varphi, \psi \in H^{1,2}(B; I\!\!R^3)$ and suppose $Z \in H_o^{1,2}(B; I\!\!R^3)$ weakly solves the equation

$$(5.1) \qquad \triangle Z = \varphi_u \wedge \psi_v + \psi_u \wedge \varphi_v \quad \text{in } B.$$

Then Z is continuous on $\overline{B}$.

Corollary 5.2:   Any weak solution $X \in \mathcal{C}(\Gamma)$ to (1.1)-(1.3) is continuous on $\overline{B}$.

Proof: Decompose $X = X_o + Z$, where $X_o \in \mathcal{C}_o(\Gamma)$,   i.e. $\triangle X_o = 0$,   and $Z \in H_o^{1,2}(B; I\!\!R^3)$. By the maximum principle and (1.3) X_o is continuous. Now Z weakly solves

$$\triangle Z = 2H \, X_u \wedge X_v \quad \text{in } B$$

and therefore is continuous by Theorem 5.1 .

$\square$

The proof of Theorem 5.1 will be a consequence of the following

Lemma 5.3:   Under the assumptions of Theorem 5.1 we have $Z \in L^\infty(B; I\!\!R^3)$ and

$$||Z||_{L^\infty} \leq 2 \, (D(\varphi)D(\psi))^{1/2} \ .$$

Proof: First note that (5.1) is conformally invariant. Indeed, if g is a conformal diffeomorphism of B and if $Z' = Z \circ g$, $\varphi' = \varphi \circ g$, $\psi' = \psi \circ g$, we have for any $\xi \in C_o^\infty(B; I\!\!R^3)$

$$\int_B \nabla Z \nabla \xi dw = \int_B \nabla Z' \nabla \xi' dw,$$

$$\int_B (\varphi_u \wedge \psi_v + \psi_u \wedge \varphi_v) \cdot \xi dw = \int_B (\varphi_u' \wedge \psi_v' + \psi_u' \wedge \varphi_v') \cdot \xi' dw$$

where $\xi' = \xi \circ g$.

Since for any $w \in B$ there is a conformal diffeomorphism of B mapping w into 0 it hence suffices to prove the estimate

$$(5.2) \qquad |Z(0)| \leq 2 \, (D(\varphi)D(\psi))^{1/2}$$

for solutions Z of (5.1) with a Lebesgue point at   0, i.e. such that

(5.3) $$\lim_{r \to 0} r^{-2} \int_{B_r(0)} |Z(w) - Z(0)| dw = 0.$$

Introducing polar coordinates (r, ϕ) on B, note that e.g.

$$\varphi_u = \cos \phi \, \varphi_r - (1/r) \sin \phi \, \varphi_\phi,$$
$$\varphi_v = \sin \phi \, \varphi_r + (1/r) \cos \phi \, \varphi_\phi,$$

whence (5.1) may be rewritten in the form:

$$\triangle Z = (1/r) \left(\varphi_r \wedge \psi_\phi + \psi_r \wedge \varphi_\phi \right).$$

For $\epsilon > 0$ and some unit vector $a \in I\!\!R^3$ let

$$\xi = \xi_\epsilon(r) \equiv a \cdot \inf\{\ln(1/r), \ln(1/\epsilon)\} \in H_o^{1,2} \cap C^o(\overline{B}; I\!\!R^3).$$

Approximating ξ in $H_o^{1,2} \cap C^o$ by smooth functions we can justify inserting ξ as a testing function in (5.1). This gives, cp. (1.9):

$$1/\epsilon \int_{\partial B_\epsilon(0)} Z \cdot a \, d\phi = \int_B \nabla Z \nabla \xi \, dw$$

$$= - \int_0^1 \int_0^{2\pi} (\varphi_r \wedge \psi_\phi + \psi_r \wedge \varphi_\phi) \cdot \xi d\phi \, dr$$

$$= - \int_0^1 \int_0^{2\pi} (\xi_r \wedge \psi_\phi + \psi_r \wedge \xi_\phi) \cdot \varphi d\phi \, dr$$

$$= \int_\epsilon^1 \int_0^{2\pi} (1/r) a \wedge \psi_\phi \cdot \varphi d\phi \, dr.$$

For $r \in [\epsilon, 1]$ let

$$\overline{\varphi}(r) = 1/(2\pi r) \int_{\partial B_r(0)} \varphi d\phi.$$

Then for a.e. $r \in [\epsilon, 1]$ by Hölder's and Poincare's inequalities:

$$\int_0^{2\pi} a \wedge \psi_\phi \cdot \varphi d\phi = \int_0^{2\pi} a \wedge \psi_\phi \cdot (\varphi - \overline{\varphi}) d\phi$$

$$\leq \left(\int_0^{2\pi} |\psi_\phi|^2 d\phi \int_0^{2\pi} |\varphi - \overline{\varphi}|^2 d\phi \right)^{1/2}$$

$$\leq 2\pi \left(\int_0^{2\pi} |\psi_\phi|^2 d\phi \int_0^{2\pi} |\varphi_\phi|^2 d\phi \right)^{1/2}$$

$$\leq 2\pi r \left(\int_0^{2\pi} |\nabla \psi|^2 r d\phi \int_0^{2\pi} |\nabla \varphi|^2 r \, d\phi \right)^{1/2} .$$

Hence

$$\left| (1/\epsilon) \int_{\partial B_\epsilon(0)} Z d\phi \right| \leq 2\pi \left(\int_\epsilon^1 \int_0^{2\pi} |\nabla\psi|^2 r d\phi \, dr \cdot \int_\epsilon^1 \int_0^{2\pi} |\nabla\varphi|^2 r \, d\phi \, dr \right)^{1/2}$$

$$\leq 4\pi \left(D(\psi) D(\varphi) \right)^{1/2} .$$

But by (5.3)

$$\left| 2\pi Z(0) - (1/\epsilon) \int_{\partial B_\epsilon(0)} Z \, d\phi \right| \leq (1/\epsilon) \int_{\partial B_\epsilon(0)} |Z - Z(0)| \, d\phi \to 0$$

if we let $\epsilon \to 0$ suitably, and (5.2) and the lemma follow.

$\square$

Remark 5.4: The optimal constant in Lemma 5.3 is

$$\|Z\|_{L^\infty} \leq (1/\pi) \left(D(\varphi) D(\psi) \right)^{1/2} .$$

If φ and ψ are conformal (e.g. if $\varphi = \psi = X$ where X solves (1.1) - (1.3)) one even finds the estimate

$$\|Z\|_{L^\infty} \leq (1/2\pi) \left(D(\varphi) D(\psi) \right)^{1/2} ,$$

which is best possible, cp. Wente [5].

Proof of Theorem 5.1: Introducing tangential differences

$$Z^{(h)}(r,\phi) \equiv Z(r,\phi + h) - Z(r,\phi), \quad \varphi^+(r,\phi) \equiv \varphi(r,\phi + h)$$

we have

$$\begin{aligned} \triangle Z^{(h)} = &\varphi_u^{(h)} \wedge \psi_v^{(h)} + \psi_u^{(h)} \wedge \varphi_v^{(h)} \\ &+ \varphi_u^{(h)} \wedge \psi_v + \psi_u \wedge \varphi_v^{(h)} \\ &+ \varphi_u \wedge \psi_v^{(h)} + \psi_u^{(h)} \wedge \varphi_v. \end{aligned}$$

(5.4)

By Lemma 5.3

$$\|Z^{(h)}\|_{L^\infty} \leq c \left(D(\varphi^{(h)})^{1/2} + D(\psi^{(h)})^{1/2} \right) \to 0 \ (h \to 0),$$

and Z is uniformly continuous in tangential direction.

Given $\epsilon > 0$ choose $\delta_1 > 0$ such that for any $r \in [0,1]$

$$|Z(r,\phi) - Z(r,\phi')| < \epsilon$$

for $|\phi - \phi'| < \delta_1$.

Note that for any connected interval I_{δ_1} of length δ_1

$$\underset{\phi_o \in I_{\delta_1}}{\text{ess inf}} \int_0^1 |\nabla Z|^2 r dr|_{\phi=\phi_o} \leq \delta_1^{-1} \int_{I_{\delta_1}} \int_0^1 |\nabla Z|^2 r dr \, d\phi \leq 2\delta_1^{-1} D(Z).$$

Let δ_2 be such that

$$\sqrt{4\delta_2 \delta_1^{-1} D(Z)} < \epsilon.$$

Then for any pair of points (r, ϕ), $(r', \phi') \in B$ with $\frac{1}{2} \leq r \leq r' \leq 1$ and with distance $|\phi - \phi'| < \delta_1, |r - r'| < \delta_2$, any ϕ_o with $|\phi_o - \phi|, |\phi_o - \phi'| < \delta_1$ there holds

$$\begin{aligned}|Z(r, \phi) - Z(r', \phi')| \leq & |Z(r, \phi) - Z(r, \phi_o)| \\ & + |Z(r, \phi_o) - Z(r', \phi_o)| + |Z(r', \phi_o) - Z(r', \phi')| \\ \leq & 2\epsilon + \int_r^{r'} |\nabla Z(r, \phi_o)| dr \\ \leq & 2\epsilon + \left(2\delta_2 \int_0^1 |\nabla Z|^2 r dr|_{\phi=\phi_o} \right)^{1/2}.\end{aligned}$$

Taking the infimum with respect to ϕ_o it follows that

$$|Z(r, \phi) - Z(r', \phi')| \leq 3\epsilon, \quad \text{if } |\phi - \phi'| < \delta_1, \ |r - r'| < \delta_2, \ \frac{1}{2} \leq r \leq r' \leq 1.$$

I.e. Z is continuous on the closed annulus $\{w | \frac{1}{2} \leq |w| \leq 1\}$.

Finally, consider Z in Cartesian coordinates again, and for an arbitrary unit vector $e \in I\!\!R^2$ let

$$Z^{(h)}(w) = (Z(w + he) - Z(w)).$$

Note that for $|h| < \frac{1}{4}$ $Z^{(h)}$ is defined on the ball $B_{\frac{3}{4}}(0)$ and satisfies a system like (5.4) with boundary values uniformly tending to 0:

$$\|Z^{(h)}\|_{L^\infty\left(\partial B_{\frac{3}{4}}(0)\right)} \to 0 \ (h \to 0).$$

Decomposing $Z^{(h)} = Z_1^{(h)} + Z_2^{(h)}$ with $\triangle Z_1^{(h)} = 0$ and $Z_2^{(h)} \in H_0^{1,2}(B_{\frac{3}{4}}(0))$ we conclude from Lemma 5.3 and the maximum principle that

$$\|Z^{(h)}\|_{L^\infty(B_{\frac{3}{4}}(0))} \to 0 \ (h \to 0).$$

This concludes the proof.

□

With the aid of Theorem 5.1 the proof of Theorem I.5.1 conveys to H-surfaces and we obtain the following result, cf. Hildebrandt [3, Satz 7.1]:

Theorem 5.5:   Let $\Gamma \subset I\!R^3$ be a Jordan curve of class $C^{m,\alpha}$, $m \geq 2$, $0 < \alpha < 1$, and suppose $X \in \mathcal{C}(\Gamma)$ weakly solves (1.1)-(1.3) for some $H \in I\!R$, then $X \in C^{m,\alpha}(\overline{B}; I\!R^3)$.

In general, we do not obtain a-priori bounds. However, for "small" solutions the following variant due to Heinz [3] of the isoperimetric inequality Theorem I.4.9 in conjunction with Lemma I.4.3 allows to formulate a-priori estimates for normalized H-surfaces $X \in C^*(\Gamma)$ analogous to Theorem I.5.1.

Theorem 5.6:   Let $\Gamma \subset B_R(0) \subset I\!R^3$ be a rectificiable Jordan curve of length $L(\Gamma)$,   and let $H \in I\!R$ satisfy $|H|R < 1$. Then for any H-surface $X \in \mathcal{C}(\Gamma)$ with $\|X\|_{L^\infty} \leq \frac{1}{|H|}$ there holds the estimate

$$4\pi D(X) \leq \frac{1 + |H|R}{1 - |H|R} L(\Gamma)^2.$$

We prove the qualitative estimate for smooth Γ.

Proof: Note that as in the proof of Theorem 3.1, by the maximum principle for sub-harmonic functions there holds $\|X\|_{L^\infty} \leq R$ for any H-surface $X \in \mathcal{C}(\Gamma)$ with $\|X\|_{L^\infty} \leq \frac{1}{|H|}$.

But then by (3.4)

$$0 = \int_B [-\triangle X + 2H X_u \wedge X_v] \cdot X \, dw$$

$$= \int_B |\nabla X|^2 + 2H X_u \wedge X_v \cdot X \, dw - \int_{\partial B} \partial_n X \cdot X \, do$$

$$\geq 2(1 - |H|R)D(X) - \|X\|_{L^\infty} \int_{\partial B} |\partial_n X| do.$$

Since by (1.2) and Theorem 5.5

$$\int_{\partial B} |\partial_n X| do = \int_{\partial B} \left| \frac{\partial}{\partial \phi} X \right| do = L(\Gamma)$$

the claim follows.

□

Although for H-surfaces in general no a-priori bounds on the area or their modulus exist, some partial results are known. The next result is due to Wente [5]:

Theorem 5.7:   Suppose $\Gamma \subset B_R(0) \subset I\!\!R^3$ is a Jordan curve, $X \in \mathcal{C}(\Gamma)$ an H–surface spanning Γ. Then

$$\|X\|_{L^\infty} \leq R + (1/2\pi)|H|D(X).$$

The bound cannot be improved.

For embedded H–surfaces Serrin [1] states the following:

Theorem 5.8:   Suppose $X \in \mathcal{C}(\Gamma)$ is an embedded H–surface spanning a Jordan curve $\Gamma \subset B_R(0) \subset I\!\!R^3$. Then

$$\|X\|_{L^\infty} \leq R + 2/|H|.$$

The bound cannot be improved.

Serrin's proof is based on the Alexandrov reflection principle and a-priori estimates for the non-parametric mean curvature equation (describing H–surfaces which can be represented as graphs over $I\!\!R^2$). However, there appear to be gaps in his argument which still wait to be filled. (The surface X might wind around and "reenter" through the curve Γ in such a way that upon reflecting a portion of X in a plane this reflected surface might touch X with the "wrong" orientation for the maximum principle to be applicable. This is impossible for closed surfaces.)

Remark that from the surfaces discovered by Wente [6] one can construct immersed surfaces of genus 1 having constant mean curvature but which do not satisfy the bound of Theorem 5.8.

IV. Unstable H-surfaces

1. H - extensions. In the analysis of unstable minimal surfaces we relied on the existence of harmonic extensions of admissible parametrizations of Γ in order to reformulate the Plateau problem in terms of a variational problem on a convex set. To imitate this procedure for H-surfaces we now consider *Dirichlet's problem for the H-surface system:*

Given $X_o \in H^{1,2} \cap L^\infty(B; I\!\!R^3)$, $H \in I\!\!R$ find $X \in H^{1,2} \cap C^2(B; I\!\!R^3)$ such that

(1.1) $$\triangle X = 2H X_u \wedge X_v \text{ in } B,$$
(1.2) $$X = X_o \text{ on } \partial B$$

or, equivalently, find $X \in X_o + H_o^{1,2}(B; I\!\!R^3)$ such that

$$dD_H(X) = 0 \in \left(H_o^{1,2}(B; I\!\!R^3)\right)^*.$$

Recall that for $H = 0$ the harmonic extension $\underline{X}$ of $X_o \in H^{1,2} \cap L^\infty(B; I\!\!R^3)$ is uniquely characterized by the relations

(1.3) $$\underline{X} \in X_o + H_o^{1,2}(B; I\!\!R^3),$$
(1.4) $$D(\underline{X}) = \inf\{D(X) | X \in X_o + H_o^{1,2}(B; I\!\!R^3)\}.$$

Moreover, $\underline{X}$ depends differentiably on X_o in $H^{1,2} \cap L^\infty$ by linearity of the Poisson equation, the variational characterization (1.4), and the maximum principle.

Due to the nonlinear character of the H-surface system (1.1) for $H \neq 0$ certain smallness conditions will have to be satified. Under hypotheses similar to those of Theorem III.3.1, however, we can establish the existence of $H-$ extension operators with properties analogous to the harmonic extension operator. Moreover, $H-$ extensions can be characterized by a variational principle consistent with (1.3)-(1.4).

The following result again is due to Hildebrandt [2]:

Theorem 1.1: Let $X_o \in H^{1,2} \cap L^\infty(B; I\!\!R^3)$, $H \in I\!\!R$. Let $R = \|X_o\|_{L^\infty}$, and suppose that $|H|R \leq 1$. Then there exists a solution X_H of (1.1), (1.2) characterized by the conditions that

(1.5) $\|X_H\|_{L^\infty} \leq R$,

(1.6) $D_H(X_H) = \inf\left\{D_H(X) \mid X \in X_o + H_o^{1,2}(B; I\!\!R^3), \|X\|_{L^\infty} \leq \dfrac{1}{|H|}\right\}.$

If $|H|R < 1$ then X_H is a relative minimizer of D_H on the affine space $X_o + H_o^{1,2} \cap L^\infty(B; I\!\!R^3)$.

Proof: By Lemmata III.3.2, III.3.3 D_H is coercive and weakly lower semi-continuous (with respect to the $H^{1,2}$–topology) on the space $\mathcal{S}_H := \{X \in X_o + H_o^{1,2}(B; I\!\!R^3)| \ \|X\|_{L^\infty} \le 1/|H|\}$. Hence D_H achieves its infimum at $X_H \in \mathcal{S}_H$.

If $|H|R < 1$ as in the proof of Theorem III.3.1 $\|X_H\| \le R$ and X_H yields a relative minimum of D_H on $X_o + H_o^{1,2} \cap L^\infty(B; I\!\!R^3)$. In particular,

$$dD_H(X_H) = 0 \in \left(C_o^\infty(B; I\!\!R^3)\right)',$$

whence by density of $C_o^\infty(B; I\!\!R^3)$ in $H_o^{1,2}(B; I\!\!R^3)$ and Remark III.2.4.ii) X_H solves (1.1), (1.2).

The case $|H|R = 1$ is treated as in Theorem III.3.1.

□

In order to establish continuous dependance of the solution X_H of (1.5), (1.6) on the data X_o we now investigate the behavior of D_H near a relative minimizer in $X_o + H_o^{1,2} \cap L^\infty(B; I\!\!R^3)$. The following result is due to Brezis-Coron [1] :

Lemma 1.2: If $X \in X_o + H_o^{1,2}(B; I\!\!R^3)$ is a relative minimizer of D_H with respect to $H_o^{1,2} \cap L^\infty(B; I\!\!R^3)$ then X is a strict relative minimizer of D_H on $X_o + H_o^{1,2}(B; I\!\!R^3)$ and for some $\delta > 0$ there holds

$$d^2 D_H(X)(\varphi, \varphi) \ge \delta D(\varphi), \quad \forall \varphi \in H_o^{1,2}(B; I\!\!R^3).$$

Proof: By density of C_o^∞ in $H_o^{1,2}$ clearly

(1.7) $d^2 D_H(X)(\varphi, \varphi) \ge 0$ for all $\varphi \in H_o^{1,2}(B; I\!\!R^3)$.

Let

$$\delta = \inf \left\{ d^2 D_H(X)(\varphi, \varphi)| \ \varphi \in H_o^{1,2}(B; I\!\!R^3), \ D(\varphi) = 1 \right\} \ge 0.$$

The lemma follows once we establish that $\delta > 0$. Suppose by contradiction that $\delta = 0$. Let $\varphi_m \in H_o^{1,2}(B; I\!\!R^3)$ be a sequence such that $D(\varphi_m) = 1$ while

$$d^2 D_H(X)(\varphi_m, \varphi_m) \to 0 \quad (m \to \infty).$$

We may assume that $\varphi_m \xrightarrow{w} \varphi$ weakly in $H_o^{1,2}(B; I\!\!R^3)$ whence by Theorem III.2.3 and Example I.3.4

$$0 \le d^2 D_H(X)(\varphi, \varphi) = 2D(\varphi) + 2H \, d^2 V(X)(\varphi, \varphi)$$

$$\le \liminf_{m \to \infty} \left(2D(\varphi_m) + 2H d^2 V(X)(\varphi_m, \varphi_m)\right)$$

$$= \liminf_{m \to \infty} d^2 D_H(X)(\varphi_m, \varphi_m) = 0.$$

Since by Theorem III.2.3

$$d^2V(X)(\varphi_m, \varphi_m) \to d^2V(X)(\varphi, \varphi) \quad (m \to 0)$$

it follows that $D(\varphi_m) \to D(\varphi)$, whence $\varphi_m \to \varphi$ strongly in $H_o^{1,2}(B; I\!R^3)$, $D(\varphi) = 1$, and

(1.8) $$d^2D_H(X)(\varphi, \varphi) = 0.$$

By (1.7) φ minimizes the functional $E(\psi) \equiv d^2D_H(X)(\psi, \psi)$ on $H_o^{1,2}(B; I\!R^3)$. Necessarily then for all $\psi \in H_o^{1,2}(B; I\!R^3)$

$$d^2D_H(X)(\varphi, \psi) = 0,$$

i.e. $\varphi \in H_o^{1,2}(B; I\!R^3)$ weakly solves the equation

$$\triangle\varphi = 2H\left(X_u \wedge \varphi_v + \varphi_u \wedge X_v\right) \quad \text{in } B.$$

From Lemma III.5.2 we infer that $\varphi \in L^\infty(B; I\!R^3)$. X being relatively minimal for D_H on $X_o + H_o^{1,2} \cap L^\infty(B; I\!R^3)$ we now obtain from Remark III.2.4.i) and (1.8) that for $|t| << 1$:

$$D_H(X) \le D_H(X + t\varphi) = D_H(X) + 2Ht^3V(\varphi).$$

I.e.

$$V(\varphi) = 0, \ D_H(X + t\varphi) \equiv D_H(X),$$

and all functions $X + t\varphi$, $|t| << 1$, must be relative minima of D_H with respect to $H_o^{1,2} \cap L^\infty(B; I\!R^3)$ and hence solutions to (1.1), (1.2):

$$\triangle(X + t\varphi) = 2H(X + t\varphi)_u \wedge (X + t\varphi)_v \quad \text{in } B.$$

Differentiating twice with respect to t we obtain that

$$\varphi_u \wedge \varphi_v \equiv 0 \quad \text{in } B,$$

whence

$$d^2D_H(X)(\varphi, \varphi) = 2D(\varphi) = 2.$$

But this contradicts (1.8) and hence we must have $\delta > 0$.

□

Corollary 1.3:   If $X_1, X_2 \in X_o + H_o^{1,2}(B; I\!R^3)$ are relative minimizers of D_H on $X_o + H_o^{1,2}(B; I\!R^3)$, with respect to variations in $H_o^{1,2} \cap L^\infty(B; I\!R^3)$, then $X_1 = X_2$.

Proof: Let $\varphi = X_1 - X_2$. Adding the expansions

$$D_H(X_1) = D_H(X_2 + \varphi) = D_H(X_2) + 1/2 \, d^2D_H(X_2)(\varphi, \varphi) + 2H \, V(\varphi)$$
$$D_H(X_2) = D_H(X_1 - \varphi) = D_H(X_1) + 1/2 \, d^2D_H(X_1)(\varphi, \varphi) - 2H \, V(\varphi)$$

we obtain that

$$0 = d^2 D_H(X_1)(\varphi, \varphi) + d^2 D_H(X_2)(\varphi, \varphi) \geq \delta D(\varphi),$$

by Lemma 1.2. Hence $\varphi = 0, X_1 = X_2$.

$\Box$

Corollary 1.4:   Let $H, R \in I\!R$ satisfy $|H|R < 1$. There is a unique differentiable and bounded map η_H of the space $\{X_o \in H^{1,2} \cap L^\infty(B; I\!R^3) \mid \|X_o\|_{L^\infty} \leq R\}$ into itself which associates to any X_o the unique solution X_H to (1.1), (1.2) characterized by (1.5), (1.6).

Proof: Existence and uniqueness of $X_H = \eta_H(X_o) = X_o + Z_o$, $Z_o = Z_o(X_o) \in H_o^{1,2}(B; I\!R^3)$ follows from Theorem 1.1, resp. Corollary 1.3.

By Lemma 1.2 and the implicit function theorem, for any such $X_H = X_o + Z_o$ there exists a unique local solution $Z(X)$ of the equation

$$dD_H(X + Z(X)) = 0 \in \left(H^{1,2}(B; I\!R^3)\right)^*.$$

for X in an $H^{1,2} \cap L^\infty(B; I\!R^3)$-neighborhood of X_o.

Indeed, the map $f : H^{1,2} \cap L^\infty(B; I\!R^3) \times H_o^{1,2}(B; I\!R^3) \to H_o^{1,2}(B; I\!R^3)$ given by

$$f(X, Z) = dD_H(X + Z)^* \in H_o^{1,2}(B; I\!R^3)$$

is analytic in a neighborhood of (X_o, Z_o), and the estimate

$$\left(\frac{d}{dZ} f(X_o, Z_o) \cdot \varphi, \varphi\right)_{H_o^{1,2}(B; I\!R^3)} = d^2 D_H(X_H)(\varphi, \varphi) \geq \delta \, D(\varphi)$$

following from Lemma 1.2 shows that $\frac{\partial}{\partial Z} f(X_o, Z_o)$ is an isomorphism of $H_o^{1,2}(B; I\!R^3)$. Letting $\bar{\eta}_H(X) = X + Z(X)$, by Lemma 1.2 and continuous dependance of $d^2 D_H(X)$ on $X \in H^{1,2}(B; I\!R^3)$, $\bar{\eta}_H(X)$ is a relative minimizer of D_H on $X + H_o^{1,2}(B; I\!R^3)$. Hence by Corollary 1.3 $\bar{\eta}_H = \eta_H$, and η_H is a smooth map from $H^{1,2} \cap L^\infty(B; I\!R^3)$ into $H^{1,2}(B; I\!R^3)$. Smoothness of η_H into $L^\infty(B; I\!R^3)$ follows from Lemma III.5.3:

Let $X_H = X_o + Z_o$, $\triangle X_o = 0$, $Z_o \in H_o^{1,2}(B; I\!R^3)$. Then

$$\triangle Z_o = 2H X_{H_u} \wedge X_{H_v} \quad \text{in } B,$$

and by Lemma III.5.3 the operator $\triangle^{-1}$ (with homogeneous Dirichlet data) composed with the map

$$X \mapsto X_u \wedge X_v$$

is smooth from $H^{1,2}(B; I\!R^3)$ into $L^\infty(B; I\!R^3)$.

Finally, boundedness of η_H is immediate fron (1.5), (III.3.5).

□

Remark 1.5:   i) Analogous to Theorem III.3.4 one also obtains relative minima of D_H on $X_o + H_o^{1,2}(B;I\!R^3)$ provided the condition

$$(1.7) \qquad\qquad H^2 D(X_o) < 2\pi/3$$

is satisfied for H, cp. Steffen [3] . Since (1.7), however, cannot hold simultaneously for *all* $X \in C_o(\Gamma)$ for any rectificiable Jordan curve in $I\!R^3$, this condition is not suited for constructing $H-$extensions on the whole class $C_o(\Gamma)$.

ii)   By a result of Jäger [1] actually for $|H|R < 1$ the solution $X_H = \eta_H(X_o)$ is the only solution to (1.1), (1.2) with $\|X_H\|_\infty \leq R$.

2. Ljusternik - Schnirelman and Morse theory for "small" H-surfaces.

In analogy with Section II.2, II.4 we now develop a critical point theory for "small" solutions to the Plateau problem for surfaces of prescribed mean curvature H, i.e. for solutions X to (III.1.1) -(III.1.3) spanning a Jordan curve $\Gamma \in B_R(0) \subset I\!\!R^3$ that satisfy the extra condition $\|X\|_{L^\infty} \leq R$.

Recall the definitions of the sets M (resp. $M^\dagger$) of monotone reparametrizations of $[0, 2\pi] \hat{=} \partial B$ (normalized with respect to the conformal group action). For brevity, again denote $T := H^{1/2 ,2} \cap C^o(I\!\!R/2\pi)$ with norm $|\cdot|$. Let $\gamma \in C^r$, $r \geq 2$, be a reference parametrization of Γ and let

$$X_o : M \to C_o(\Gamma) = \{X \in C(\Gamma) \,|\, \triangle X = 0\}$$

be the map

$$x \mapsto X_o(x) = h(\gamma \circ x),$$

where h denotes harmonic extension. By Lemma II.2.5 and the maximum principle the map X_o extends to a map of class C^{r-1} of $\{\,\text{id}\,\} + T$ into

$$\{X \in H^{1,2} \cap L^\infty(B; I\!\!R^3) \,|\, \|X\|_{L^\infty} \leq R\}.$$

Now suppose that

$$|H|R < 1.$$

Composing the map X_o with the H−extension map η_H constructed in Corollary 1.4 we obtain a map

$$X_H : M \to C(\Gamma)$$
$$x \mapsto X_H(x) = \eta_H(X_o(x))$$

whose image lies in the set

$$C_H(\Gamma) = \{X \in C(\Gamma) \,|\, \|X\|_{L^\infty} \leq R;\ \triangle X = 2H X_u \wedge X_v\}.$$

Moreover, we may let

$$E_H(x) = D_H(X_H(x)).$$

Lemma 2.1:　　If $\gamma \in C^r$, $r \geq 2$, X_H extends to a map of the affine space $\{\,\text{id}\,\} + T$ into $H^{1,2} \cap L^\infty(B; I\!\!R^3)$ of class C^{r-1}.

Proof:　　By smoothness of η_H X_H inherits the properties of X_o.

□

Lemma 2.2:　　If $\gamma \in C^r, r \geq 2$, the functional E_H extends to a C^{r-1}-functional on $\{\,\text{id}\,\} + T$ with derivative at $X_H = X_H(x)$ for $x \in M$ given by

$$\langle dE_H(x), \xi \rangle = \int\limits_{\partial B} \partial_n X_H \cdot \frac{d}{d\phi}\gamma(x) \cdot \xi \, do$$

for all $\xi \in T$.

Subsets of $M^\dagger$ where E_H is uniformly bounded are bounded in $H^{1/2,2}$ and relatively compact with respect to the C^o- topology.

Proof: The last statement follows from Lemma III.3.2, Lemma II.2.8, and Remark II.2.4. Moreover, by Lemma 2.1 and analyticity of D_H the assertion that $E_H \in C^{r-1}$ is obvious.

Compute for smooth x , using the fact that $X_H = X_o$ on ∂B implies that $dX_H(x) \cdot \xi = dX_o(x) \cdot \xi = \frac{d}{d\phi}\gamma(x) \cdot \xi$ on ∂B :

$$\langle dE_H(x), \xi \rangle = \langle dD_H(X_H), \, dX_H(x) \cdot \xi \rangle =$$

$$= \int_B \nabla X_H \nabla(dX_H(x) \cdot \xi)dw \, + \, (2H/3)\int_B X_{H_u} \wedge X_{H_v} \cdot (dX_H(x) \cdot \xi) +$$

$$+ \, (X_{H_u} \wedge (dX_H(x) \cdot \xi)_v + (dX_H(x) \cdot \xi)_u \wedge X_{H_v}) \cdot X_H \, dw$$

$$= \int_B [-\Delta X + 2H X_{H_u} \wedge X_{H_v}] \cdot (dX_H(x) \cdot \xi) \, dw$$

$$+ \int_{\partial B} \partial_n X \cdot \frac{d}{d\phi}\gamma(x) \cdot \xi \, do$$

$$+ (2H/3) \int_{\partial B} (v \cdot X_{H_u} - u X_{H_v}) \wedge \left(\frac{d}{d\phi}\gamma(x) \cdot \xi\right) \cdot X_H \, do.$$

Now the first and the last term on the right vanish since by definition $\Delta X_H = 2H X_{H_u} \wedge X_{H_v}$ in B, while $v \cdot X_{H_u} - u X_{H_v} = -\frac{d}{d\phi}X_H$ is co-linear with $\frac{d}{d\phi}\gamma(x)$ along ∂B.

The general case follows by density of smooth monotone maps in M . Note that all integrals exist in the distribution sense. In particular, since X_H solves (III.1.1) the estimate for harmonic $\varphi \in H^{1,2} \cap L^\infty(B; \mathbb{R}^3)$

$$\int_{\partial B} \partial_n X_H \cdot \varphi \, do = \int_B \nabla X_H \nabla \varphi dw + 2H \int_B X_{H_u} \wedge X_{H_v} \cdot \varphi dw$$

$$\leq cD(X_H)^{1/2} D(\varphi)^{1/2} + D(X_H)|H| \, \|\varphi\|_{L^\infty}$$

$$\leq cD(X_H)^{1/2} |\varphi|_{H^{1/2,2}(\partial B)} + D(X_H)|H| \, \|\varphi\|_{L^\infty}$$

shows that $\partial_n X_H$ exists as a distribution on $H^{1/2,2} \cap L^\infty(\partial B; \mathbb{R}^3)$.

$\Box$

Define

$$g_H(x) = \sup_{\substack{y \in M^\dagger \\ |x-y| < 1}} \langle dE_H(x), \, x - y \rangle.$$

Analogous to Proposition II.2.9 one can show:

Lemma 2.3: $x \in M^\dagger$ satisfies $g_H(x) = 0$ iff $X = X_H(x)$ solves (III.1.1)–(III.1.3).

Moreover, E_H satisfies the Palais-Smale condition on $M^\dagger$:

Lemma 2.4:   Any sequence $\{x_m\} \subset M^\dagger$ such that $|E_H(x_m)| \leq c$ uniformly while $g_H(x_m) \to 0$ $(m \to \infty)$ is relatively compact.

Proof: Let $X_m = X_H(x_m) = X_{mo} + Z_m$, where $X_{mo} \in \mathcal{C}_o(\Gamma)$, $Z_m \in H_o^{1,2}(B; \mathbb{R}^3)$.

By Lemma 2.2 we may assume that for some $x \in M^\dagger$ $x_m \xrightarrow{w} x$ weakly in $H^{1/2,2}$ and uniformly, whence also $X_m \xrightarrow{w} X$ weakly in $H^{1,2}(B; \mathbb{R}^3)$ and $X_{mo} \to X_o$ uniformly on $\overline{B}$. As in the proof of Lemma II.2.12

$$X_{mo} - X_o = \frac{d}{d\phi}\gamma(x_m)(x_m - x) + I_m \quad \text{on } \partial B \,,$$

where $I_m \to 0$ strongly in $H^{1/2,2} \cap C^o(\partial B)$.

Hence

$$o(1) \geq g_H(x_m)|x_m - x| \geq \int_{\partial B} \partial_n X_m \cdot \frac{d}{d\phi}\gamma(x_m)(x_m - x) \, do$$

$$= \int_{\partial B} \partial_n X_m \cdot (X_{mo} - X_o) \, do + o(1)$$

$$= \int_B \nabla X_m \, \nabla(X_{mo} - X_o) \, dw + 2H \int_B X_{m_u} \wedge X_{m_v} \cdot (X_{mo} - X_o) \, dw + o(1)$$

$$= \int_B \nabla(X_m - X) \, \nabla(X_{mo} - X_o) \, do + o(1).$$

By orthogonality

$$\int_B \nabla X \nabla Z \, dw = 0, \quad \forall \, X \in \mathcal{C}_o(\Gamma), \, Z \in H_o^{1,2}(B; \mathbb{R}^3)$$

the latter implies that

$$\int_B |\nabla(X_{mo} - X_o)|^2 dw \leq o(1).$$

I.e. $X_{mo} \to X_o$ in $H^{1,2}(B; \mathbb{R}^3)$, and $x_m \to x$ in $H^{1/2,2} \cap C^o$ as claimed.

$\square$

Applying Theorem II.1.13 we hence obtain the following result which strengthens results of Heinz [5] and Ströhmer [1] on the existence of "small" unstable H-surfaces in $\mathcal{C}_H(\Gamma)$:

Theorem 2.5:   Let $\Gamma \subset B_R(0) \subset \mathbb{R}^3$ be a Jordan curve of class C^2, $H \in \mathbb{R}$, $|H|R < 1$. Assume D_H admits two distinct relative minima $X_1, X_2 \in \mathcal{C}_H(\Gamma)$. Then either there exists an unstable solution $X_3 \in \mathcal{C}_H(\Gamma)$ to the Plateau problem (III.1.1) - (III.1.3) or $D_H(X_1) = D_H(X_2) = \beta$ and X_1, X_2 can be connected in any neighborhood of the set of relative minima X of D_H in $\mathcal{C}_H(\Gamma)$ with $D_H(X) = \beta$.

In order to extend the Morse theory of Section II.4 to "small" solutions to the Plateau problem for surfaces of constant mean curvature note the following

Lemma 2.6:   Suppose $\Gamma \subset B_R(0) \subset \mathbb{R}^3$ is of class C^5,   with reference parametrization γ, and $H \in \mathbb{R}$ satisfies $|H|R < 1$.

Then the functional $E_H \in C^3(\{\,\mathrm{id}\,\} + T)$,   and at any critical point $x_o \in M^\dagger$ (where $g_H(x_o) = 0$) the expansions hold:

$$E_H(x) = E_H(x_o) + 1/2\, d^2 E_H(x_o)(x - x_o, x - x_o) + o(|x - x_o|_{1/2}^2),$$

$$\langle dE_H(x), x - y \rangle = d^2 E_H(x_o)(x - x_o, x - x_o) + o(|x - x_o|_{1/2}^2),$$

for all $x, y \in M^\dagger$,   such that $|x - y|_{1/2} \leq |x - x_o|_{1/2}$

The **proof** is the same as that of Lemma II.4.2.

Now let

$$T^\dagger = \left\{ \xi \in T \,\Big|\, \int_0^{2\pi} \xi \cdot \eta \, d\phi = 0, \; \forall \eta \in T_{id}G \right\}$$

$$H^\dagger = H^{1/2,2} - clos(T^\dagger)$$

as in Section II.4 and consider

$$M^\dagger \subset \{\,\mathrm{id}\,\} + T^\dagger \subset \{\,\mathrm{id}\,\} + H^\dagger.$$

Lemma 2.7:   Under the hypotheses of Lemma 2.6 if $x_o \in M^\dagger$ is critical, the bilinear form $d^2 E_H(x_o)$ extends to a bilinear form on $H^\dagger$ and induces a decomposition $H^\dagger = H_+^\dagger \oplus H_o^\dagger \oplus H_-^\dagger$. Moreover, $dim\left(H_o^\dagger \oplus H_-^\dagger\right) < \infty$ and $H_o^\dagger \oplus H_-^\dagger \subset C^1(\mathbb{R}/2\pi)$.

The **proof** of Proposition II.5.6 conveys with minor modifications.

As before a critical point $x_o \in M^\dagger$ will be called non-degenerate if $d^2 E_H(x_o)$ induces an isomorphism of $H^\dagger$.

Lemma 2.8:　　Under the hypotheses of Lemma 2.6, if　$x_o \in M^\dagger$　is a non-degenerate critical point of　E_H　and　$H^\dagger = H_+^\dagger \oplus H_-^\dagger$　denotes the standard decomposition of　$H^\dagger$　at　x_o, then there exists a neighborhood　u_-　of　0　in　$H_-^\dagger$　such that

$$x_0 + U_- \subset M^\dagger.$$

Proof:　　x_o　induces a regular solution　$X = X_H(x_o)$　of (III.1.1)-(III.1.3). Since branch points of　X　give rise to forced Jacobi fields　$\xi \in H^\dagger$　with

$$d^2 E_H(x_o)(\xi, \eta) = 0, \quad \forall \eta \in H^\dagger$$

(cp. Söllner [1]) our non-degeneracy assumption implies that　X　is immersed over　$\overline{B}$. In particular,　$\frac{d}{d\phi} x_o \geq c > 0$, and the claim follows from Lemma 2.7.

□

By Lemma 2.6 and Lemma 2.8 Theorem II.3.6 applies to　E_H　on　$M^\dagger$　and we obtain the following result:

Theorem 2.9:　　Suppose　$\Gamma \subset B_R(0) \subset I\!R^3$　is a Jordan curve parametrized by a diffeomorphism　$\gamma \in C^5$, and　$H \in I\!R$　satisfies the bound　$|H| R < 1$. Suppose that all small　H−surfaces　$X = X_H(x) \in \mathcal{C}_H(\Gamma)$　bounded by　Γ　correspond to non-degenerate critical points　x　of　E_H　in　$M^\dagger$. Then the Morse inequalities (II.3.4) hold.

Remark 2.10:　　As in the case of minimal surfaces the non-degeneracy of a parametrization　x　of an　H−surface　$X = X_H(x) \in \mathcal{C}_H(\Gamma)$　as a critical point of E_H　will be equivalent to the non-degeneracy of　X　on the space　$dX_H(x)(H^\dagger)$　of surfaces "tangent" to　$\mathcal{C}_H(\Gamma)$　at　X　and the Morse indeces correspond, cp. Remark II.4.9.

3. Large solutions to the Dirichlet problem. By Corollary 1.3 solutions to (1.1)-(1.2) of minimum type are unique. However, any time there exists a relative minimum of D_H on $X_o + H_o^{1,2}(B; I\!R^3)$ for $H \neq 0$ the global behavior of D_H as a cubic functional lets us expect a further solution which is not of minimum type.

The following result is due to Brezis-Coron [1] and the author [3] with an extension by Steffen [4]. More precisely, in Struwe [3] the thesis was obtained for "admissible" $X_o \in H^{1,2} \cap L^\infty(B; I\!R^3)$ and small $H \neq 0$, while Steffen was able to show that all non-constant X_o are "admissible" in that sense. Independently and almost simultaneously Brezis and Coron found non-uniqueness for non-constant $X_o \in H^{1,2} \cap L^\infty(B; I\!R^3)$ and $H \in I\!R$ satisfying the bounds $0 < |H|R < 1$, where $R = \|X_o\|_{L^\infty}$. Moreover, their proof extends to the general case considered below.

Theorem 3.1: Let $X_o \in H^{1,2} \cap L^\infty(B; I\!R^3)$, $H \in I\!R$, and suppose that $X_o \not\equiv$ const, $H \neq 0$, and that D_H admits a relative minimum X_H on $X_o + H_o^{1,2}(B; I\!R^3)$. Then there also exists an unstable solution X^H of (1.1)-(1.2) on $X_o + H_o^{1,2}(B; I\!R^3)$.

Theorem 3.1 in particular applies if $0 < |H| \|X_0\|_{L^\infty} < 1$ or $H^2 D(X_o) < 2\pi/3$. The result is optimal in the sense that if $X_o \equiv$ const or $H = 0$ the solution to (1.1)-(1.2) is unique. This is a consequence of a result due to Wente [4] :

Theorem 3.2: Suppose $X \in H_o^{1,2}(B; I\!R^3)$ satisfies (1.1)-(1.2) for some $H \in I\!R$ with boundary data $X_o \equiv 0$. Then $X \equiv 0$.

Proof: Extending X by reflection

$$\overline{X}(w) = \begin{cases} X(w), & |w| \leq 1 \\ -X\left(\dfrac{w}{|w|^2}\right), & |w| > 1, \end{cases}$$

we obtain a continuous weak solution $\overline{X}$ to (1.1) in $I\!R^2$ with

$$D(\overline{X}; I\!R^2) = 2D(X) < \infty.$$

Letting $F(w) = \overline{X}_u - i\overline{X}_v$ we hence obtain that F is a holomorphic function with
$$F^2 = |\overline{X}_u|^2 - |\overline{X}_v|^2 - 2i\overline{X}_u \cdot \overline{X}_v \in L^1(I\!R^2).$$

By the mean value theorem $F \equiv 0$ and $\overline{X}$ is conformal. But then $\overline{X}$ can have only finitely many branch points on ∂B or $\overline{X} \equiv$ const $= 0$. Since $\overline{X} \equiv 0$ on ∂B, by conformality also $\nabla \overline{X} \equiv 0$ on ∂B, and the conclusion follows.

□

As in Struwe [2] and consistent with the remainder of this book Theorem 3.1 will now be deduced as an application of the Mountain - Pass - Lemma in the following variant (cf. Theorem II.1.12):

Theorem 3.3:   Let T be an (affine) Banach space, $E \in C^1(T)$ and suppose E admits a relative minimum $\underline{x}$ and a point x_1 where $E(x_1) < E(\underline{x})$.

Define

(3.1)
$$P = \{p \in C^o([0,1]; T) \mid p(0) = \underline{x}, \ p(1) = x_1\}$$

and let

(3.2)
$$\beta = \inf_{p \in P} \sup_{x \in p} E(x).$$

Assume that E satisfies the Palais - Smale condition at level β, i.e. the condition:

$(P.S.)_\beta$ Any sequence $\{x_m\}$ in T such that $E(x_m) \to \beta$ while $dE(x_m) \to 0$ as $m \to \infty$ is relatively compact.

Then E admits an unstable critical point $\overline{x}$ with $E(\overline{x}) = \beta$.

Proof of Theorem 3.3:   Note that Lemma II.1.10 with $M = T$ remains true at the level β under the weaker compactness condition $(P.S.)_\beta$. But then also Lemma II.1.9 remains true at level β, and the proof of Theorem II.1.12 conveys: For $\overline{\epsilon} = E(\underline{x}) - E(x_1) > 0$ and any neighborhood N of the set K_β of critical points x of E with $E(x) = \beta$ there exists a number $\epsilon \in]0, \overline{\epsilon}[$ and a flow Φ having the properties i), ii), iii) of Lemma II.1.9. Suppose by contradiction that K_β consists of relative minima of E only. Then K_β is relatively open (and trivially closed, by continuity) in

$$\overline{M}_\beta = \{x \in T \mid E(x) \le \beta\},$$

Let N be a neighborhood of K_β such that $N \not\ni x_1$ and

$$N \cap M_{\beta-\epsilon} = \emptyset$$

for any $\epsilon > 0$, and let ϵ, Φ be chosen correspondingly. Select a path $p \in P$ such that

$$\sup_{x \in p} E(x) < \beta + \epsilon.$$

Then $p' = \Phi(p, 1) \in P$ by property i) of Lemma II.1.9, while by property iii)

$$p' \subset N \cup M_{\beta-\epsilon}.$$

N and $M_{\beta-\epsilon}$ being disjoint, it follows that either $p' \subset N$ or $p' \subset M_{\beta-\epsilon}$. But $x_1 \in p'$, and since $x_1 \notin N$ the first case cannot occur, while the second contradicts the definition of β. The contradiction proves the theorem.

□

We now apply Theorem 3.3 with $E = D_H$ on $T = X_o + H_o^{1,2}(B; \mathbb{R}^3)$, $\underline{x} = X_H$. The following result is essentially due to Wente [3]. In the generality stated below it was proved by Brezis-Coron [2]:

Lemma 3.4:   For any $H \neq 0$ the functional D_H admits a surface X_1 such that $D_H(X_1) < D_H(X_H)$. Moreover, letting P and β be given by (3.1), (3.2) , if $X_o \not\equiv$ const. we have:

$$\beta < D_H(X_H) + 4\pi/(3H^2).$$

Note that $4\pi/(3H^2)$ is the "energy" D_H of a sphere of radius $1/|H|$ in conformal representation.

The proof below is essentially due to Brezis-Coron [2] :

Proof: If $X_o \not\equiv$ const clearly $X_H \not\equiv$ const and at some point $w_o \in B$ we have $\nabla X_H(w_o) \neq 0$. By translation and rotation of coordinates we may assume that $w_o = 0$, and

$$\frac{\partial}{\partial u} X_H(0) = \left(a^1, a^2, a^3\right), \; \frac{\partial}{\partial v} X_H(0) = \left(b^1, b^2, b^3\right)$$

satisfy the condition that

(3.3) $H\left(a^1 + b^2\right) < 0.$

For $\epsilon > 0$ now let

$$\varphi^\epsilon(u,v) = \frac{2\epsilon}{\epsilon^2 + u^2 + v^2}\left(u, v, \epsilon\right)$$

be a conformal representation of a sphere of radius 1 around $(0,0,1)$, obtained by stereographic projection from the "north pole".

Also let $\xi \in C_o^\infty(B)$ be a symmetric cut-off function such that $\xi(w) = \xi(-w)$ and $\xi \equiv 1$ near $w = 0$.

Consider the family

$$X_t^\epsilon = X_H + t\xi\varphi^\epsilon \in X_o + H_o^{1,2}(B; I\!\!R^3).$$

For $\epsilon = 0$, X_t^ϵ can be pictured as a sphere of radius t attached to X_H at $X_H(0)$.

Compute, using (III.2.13)

$$D_H(X_t^\epsilon) = D_H(X_H) + D_H(t\xi\varphi^\epsilon) + t^2 H d^2 V(X_H)(\xi\varphi^\epsilon, \xi\varphi^\epsilon)$$

$$= D_H(X_H) + t^2 D(\xi\varphi^\epsilon) + 2t^3 H\, V(\xi\varphi^\epsilon) + 2t^2 H \int_B X_H \cdot (\xi\varphi^\epsilon)_u \wedge (\xi\varphi^\epsilon)_v \, dw.$$

Now

$$D(\xi\varphi^\epsilon) = D(\varphi^\epsilon) + 1/2 \int_B (\xi^2 - 1)|\nabla\varphi^\epsilon|^2 + 2\xi\, \nabla\xi\, \varphi^\epsilon\, \nabla\varphi^\epsilon + |\varphi^\epsilon|^2 |\nabla\xi|^2 \, dw$$

$$\leq D(\varphi^\epsilon; I\!\!R^2) + 0(\epsilon^2) = 4\pi + 0(\epsilon^2),$$

$$V(\xi\varphi^\epsilon) = V(\varphi^\epsilon) + 0(\epsilon^3)$$

$$= V(\varphi^\epsilon; I\!\!R^2) - 1/3 \int_{R^2 \backslash B} \varphi_u^\epsilon \wedge \varphi_v^\epsilon \cdot \varphi^\epsilon \, dw + 0(\epsilon^3)$$

$$= 4\pi/3 + 0(\epsilon^3).$$

Expanding $X_H(u,v) = X_H(0) + au + bv + 0(r^2)$, where $r^2 = u^2 + v^2$:

$$2H \int\limits_B X_H \cdot (\xi\varphi^\epsilon)_u \wedge (\xi\varphi^\epsilon)_v \, dw =$$

$$= 2H \int\limits_B (X_H(0) + au + bv) \cdot (\xi\varphi^\epsilon)_u \wedge (\xi\varphi^\epsilon)_v dw$$

$$+ 2H \int\limits_B 0(r^2)(\xi\varphi^\epsilon)_u \wedge (\xi\varphi^\epsilon)_v dw.$$

By (III.1.9) the first term

$$= H \int\limits_B \left(a \wedge (\xi\varphi^\epsilon)_v + (\xi\varphi^\epsilon)_u \wedge b\right) \cdot \xi\varphi^\epsilon dw$$

which by antisymmetry

$$= H \int\limits_B \left(a \wedge (0,1,0) + (1,0,0) \wedge b\right) \xi^2 \frac{4\epsilon^2}{(\epsilon^2 + r^2)^2}(u,v,\epsilon) \, dw$$

$$= H(a^1 + b^2) \int\limits_B \frac{4\epsilon^3}{(\epsilon^2 + r^2)^2}\xi^2 dw$$

$$- H \int\limits_B (a^3 u + b^3 v) \frac{4\epsilon^2}{(\epsilon^2 + r^2)^2}\xi^2 dw.$$

The last integral vanishes by symmetry of ξ. Moreover, for sufficiently small $\epsilon > 0$ we can estimate

$$\int\limits_B \frac{\epsilon^3}{(\epsilon^2 + r^2)^2}\xi^2 dw \geq \int\limits_{B_\epsilon(0)} \frac{\epsilon^3}{(\epsilon^2 + r^2)^2}\xi^2 dw$$

$$\geq c_1\epsilon > 0$$

with a uniform constant $c_1 > 0$. On the other hand since $|\nabla\varphi^\epsilon| \leq c\frac{\epsilon}{\epsilon^2 + r^2}$ we may estimate

$$\left| \int\limits_B 0(r^2) \cdot (\xi\varphi^\epsilon)_u \wedge (\xi\varphi^\epsilon)_v dw \right| \leq$$

$$\leq \left| \int\limits_B 0(r^2) \cdot |\nabla\varphi^\epsilon|^2 dw \right| + 0(\epsilon^2)$$

$$\leq c \int\limits_B \frac{\epsilon^2 r^2}{(\epsilon^2 + r^2)^2} dw + 0(\epsilon^2)$$

$$\leq c \int\limits_{B_\epsilon(0)} dw + c \int\limits_{B \setminus B_\epsilon(0)} \frac{\epsilon^2}{r^2} dw + 0(\epsilon^2)$$

$$\leq c_2 \cdot \epsilon^2 |\ln \epsilon| + 0(\epsilon^2).$$

Hence for sufficiently small $\epsilon > 0$

$$D_H(X_t^\epsilon) \le D_H(X_H) + \left(4\pi + 4H(a^1 + b^2)c_1\epsilon + c_2\epsilon^2|\ln\epsilon| + 0(\epsilon^2)\right) t^2$$
$$+ 2H\left(4\pi/3 + 0(\epsilon^3)\right) t^3.$$

Clearly if $Ht \to -\infty$ we have $D_H(X_t^\epsilon) \to -\infty$. Hence if $H\neq0$ a surface $X_1 = X_{t_1}^\epsilon$ exists as claimed. Moreover, suppose $H < 0$. Then, if $X_o \neq$ const. and if (3.3) holds, for sufficiently small $\epsilon > 0$ the value $\sup_{t\ge0} D_H(X_t^\epsilon)$ is achieved at $t_o \le 1/|H| - c_3\epsilon$, where $c_3 > 0$. Hence in this case

$$\sup_{t\ge0} D_H\left(X_t^\epsilon\right) < D_H(X_H) + 4\pi/(3H^2),$$

and the lemma follows. The case $H > 0$ may be treated similarly.

$\square$

Finally, we establish the local Palais-Smale condition $(P.S.)_\beta$.

Lemma 3.5: Let $H\neq0$ and suppose that X_H is a relative minimizer of D_H on $X_o + H_o^{1,2}(B;I\!\!R^3)$. Then any sequence $\{X_m\}$ in $X_o + H_o^{1,2}(B;I\!\!R^3)$ such that

$$D_H(X_m) \to \beta < D_H(X_H) + \frac{4\pi}{3H^2},$$
$$dD_H(X_m) \to 0 \in \left(H_o^{1,2}(B;I\!\!R^3)\right)^*$$

is relatively compact.

Proof: To show boundedness of $\{X_m\}$ in $H^{1,2}$ observe that by Lemma 1.2 there exists $\delta > 0$ such that

$$d^2 D_H(X_H)(\varphi,\varphi) \ge \delta D(\varphi), \quad \forall\, \varphi \in H_o^{1,2}(B;I\!\!R^3).$$

Now let $\varphi_m = X_m - X_H$ and expand

$$D_H(X_m) = D_H(X_H + \varphi_m) = D_H(X_H) + 1/2\, d^2 D_H(X_H)(\varphi_m,\varphi_m) + 2H\, V(\varphi_m),$$
$$\langle dD_H(X_m), \varphi_m\rangle = d^2 D_H(X_H)(\varphi_m,\varphi_m) + 6H\, V(\varphi_m) = o(1)\, D(\varphi_m)^{1/2}.$$

Subtracting three times the first line from the second there results

$$3\left(D_H(X_m) - D_H(X_H)\right) - o(1)\, D(\varphi_m)^{1/2} = 1/2\, d^2 D_H(X_H)(\varphi_m,\varphi_m) \ge \frac{1}{2}\,\delta D(\varphi_m)$$

and $D(\varphi_m) \le c$ uniformly.

Hence we may assume that $X_m \xrightarrow{w} X$ weakly in $H^{1,2}(B;I\!\!R^3)$. By weak continuity of D_H, cp. Theorem III.2.3, $dD_H(X) = 0$, whence the cubic character of D_H guarantees that

(3.4) $$D_H(X) \ge D_H(X_H).$$

Now let $\psi_m = X_m - X \xrightarrow{w} 0$ weakly in $H_o^{1,2}(B; I\!R^3)$. Expanding, using (III.2.13) and Theorem III.2.3, we obtain:

$$D_H(X_m) = D_H(X) + D_H(\psi_m) + H\, d^2 V(X)(\psi_m, \psi_m)$$

(3.5)
$$= D_H(X) + D_H(\psi_m) + o(1),$$

$$o(1) = \langle dD_H(X_m), \psi_m \rangle = \langle dD_H(\psi_m), \psi_m \rangle + 2H d^2 V(X)(\psi_m, \psi_m)$$

(3.6)
$$= 2D(\psi_m) + 6HV(\psi_m) + o(1),$$

where $o(1) \to 0$ $(m \to \infty)$.

In particular, for $m \geq m_o$ by (3.4-5) :

$$D_H(\psi_m) \leq D_H(X_m) - D_H(X) + o(1) \leq c < \frac{4\pi}{3H^2},$$

while from (3.6) we deduce that

$$3D_H(\psi_m) = 3D(\psi_m) + 6H\, V(\psi_m) = D(\psi_m) + o(1).$$

I.e. for $m \geq m_o$:

(3.7)
$$D(\psi_m) \leq c < \frac{4\pi}{H^2}.$$

But now (3.6) again and the isoperimetric inequality Theorem III.2.1 imply that

$$2D(\psi_m)\left(1 - \sqrt{\frac{H^2 D(\psi_m)}{4\pi}}\right) \leq 2D(\psi_m) + 6H\, V(\psi_m) = o(1).$$

In view of (3.7) this implies that $D(\psi_m) \to 0$ $(m \to \infty)$, and the proof is complete.

$\square$

Theorem 3.3 is now applicable, and Theorem 3.1 follows.

Remark 3.6 :　　It has been conjectured that for the Dirichlet problem (1.1)-(1.2) there will in general exist at most two distinct solutions. The following example which was kindly communicated to me by H. Wente shows that pathologies may occur if the group of symmetries of the data X_o is too large.

Example 3.7:　　Let $X_o(u,v) \equiv u$, $0 < H < 1$. By Theorem 1.1 and Remark 1.5.i) the function $X_H = X_o$ is the unique solution of (1.1), (1.2) with $\|X_H\|_{L^\infty} \leq 1$, which moreover furnishes a relative minimum of D_H on $X_o + H_o^{1,2}(B; I\!R^3)$. Theorem 3.1 now implies the existence of an unstable solution X^H of (1.1), (1.2). The image of X^H cannot lie entirely on the X^1–axis: otherwise $X_u^H \wedge X_v^H \equiv 0$ and $\triangle X^H = 0$, i.e. $\|X^H\|_{L^\infty} \leq 1$ by the maximum principle, and $X^H = X_H$. So $X^H(w)$ has a non-vanishing component in direction of the X^2– or X^3–axis at some $w \in B$. Rotating X^H around the X^1–axis hence generates a continuum of distinct solutions to (1.1), (1.2). It remains an interesting open question whether the "large" solution to (1.1), (1.2) is unique for boundary data which do not admit isometries of $I\!R^3$ as symmetries, i.e. which do not degenerate to a line segment.

4. Large solutions to the Plateau problem ("Rellich's conjecture"). A
result analogous to Theorem 3.1 also holds for the Plateau problem:

Theorem 4.1:   Let Γ be a Jordan curve of class C^2 in $I\!R^3$, $H \neq 0$, and suppose that D_H admits a relative minimum X_H on $\mathcal{C}(\Gamma)$. Then there also exists an unstable solution $X^H \in \mathcal{C}(\Gamma)$ of (III.1.1) - (III.1.3).

Remark 4.2.   i)   Theorem 4.1 for certain "admissible" curves Γ and sufficiently small $H \neq 0$ was established by the author in [3]. Steffen [4] then was able to show that in fact *all* rectifiable Jordan curves are "admissible" in the sense of Struwe [3]. Independently, and only a few weeks later Brezis and Coron [2] were able to extend their results for the Dirichlet problem and established non-uniqueness in the Plateau problem (III.1.1) - (III.1.3) for $\Gamma \subset B_R(0)$ and $0 < |H|R < 1$, a result which is optimal when Γ is a circle. Theorem 4.1 was finally established by Struwe [2].

By Theorem III.3.1, our Theorem 4.1 contains the Brezis - Coron result; moreover, Theorem 4.1 also applies in the case of Theorem III.3.4 where the method of Brezis and Coron is not applicable: If we only assume that $H^2 D(X) < \frac{2}{3}\pi$ for *some* $X \in \mathcal{C}(\Gamma)$ we cannot guarantee solvability of the Dirichlet problem (1.1), (1.2) for *all* boundary data $X_o \in \mathcal{C}(\Gamma)$, whereas Brezis and Coron crucially use the existence of H−extensions for *all* data $X_o \in \mathcal{C}(\Gamma)$.

ii)   By using results of Wente [2] on the Plateau problem with a volume constraint Steffen [1] in 1972 established the existence of large solutions to (III.1.1) - (III.1.3) for a sequence of curvatures $H_m \to 0$.

iii)   Theorem 4.1 establishes a conjecture often attributed to Rellich; however, no direct reference is known.

We now proceed to set the stage for the - rather tricky - proof of Theorem 4.1. First, however, we state the following a - priori - estimate for H−surfaces which will play a cruical role in our arguments.

Theorem 4.2:   For any H−surface X spanning a rectifiable Jordan curve $\Gamma \subset I\!R^3$ there holds the estimate

$$D(X) \leq 3D_H(X) + L(\Gamma)^2.$$

Proof: We present the proof for smooth Γ, whence $X \in C^1(\overline{B}; I\!R^3)$ by Theorem III.5.5. Simply compute, using (III.1.1) - (III.1.3):

$$0 = \int_B [-\Delta x + 2HX_u \wedge X_v] \cdot X \ dw =$$

$$= 2D(X) + 6HV(X) - \int_{\partial B} \partial_n X \cdot X \ do$$

$$\leq 3D_H(X) - D(X) + \int_{\partial B} \left| \frac{d}{d\phi} X \right| \ do \cdot \|X\|_{L^\infty(\partial B)}$$

$$\leq 3D_H(X) - D(X) + L(\Gamma) \ \|\Gamma\|_{L^\infty}.$$

Choosing the origin in $I\!R^3$ suitably, we can surely estimate

$$\|\Gamma\|_{L^\infty} \leq L(\Gamma),$$

and the theorem follows.

$\square$

Now let

$$\mathcal{M} = M^\dagger \times H_o^{1,2}(B; I\!R^3) \subset$$
$$\subset \left(\{\,\text{id}\,\} + T^\dagger \right) \times H_o^{1,2}(B; I\!R^3)$$
$$=: \mathcal{T},$$

and define a map $X : \mathcal{M} \to \mathcal{C}(\Gamma)$ by letting

(4.1) $$X(x) = X_o(x_o) + Z(z),$$

where

$$x = (x_o, z), \ x_o \in M^\dagger, \ z \in H_o^{1,2}(B; I\!R^3),$$
$$X_o(x_o) = h(\gamma \circ x_o) \in \mathcal{C}_o(\Gamma),$$
$$Z(z) = z \qquad \in H_o^{1,2}(B; I\!R^3).$$

Moreover, let

$$E_H(x) := D_H(X(x)).$$

The following lemma is immediate from Lemma II.2.5.

Lemma 4.3:   The map X extends to a differentiable map of $\mathcal{T}$ into $H^{1,2} \cap L^\infty(B; I\!R^3) + H_o^{1,2}(B; I\!R^3)$; E_H extends to a C^1–functional on $\mathcal{T}$.

As usual we define

$$g_H(x) = \sup_{\substack{y \in \mathcal{M} \\ |x-y|_{\mathcal{T}} < 1}} \langle dE_H(x), x - y \rangle$$

and call a zero of g_H a critical point of E_H on $\mathcal{M}$. We note

Lemma 4.4: $x \in \mathcal{M}$ is critical for E_H iff $X = X(x)$ solves (III.1.1) - (III.1.3).

Proof: If X solves (III.1.1) - (III.1.3) clearly $g_H(x) = 0$. Conversely, $g_H(x) = 0$ implies that

$$\langle dD_H(X), \varphi \rangle = 0, \quad \forall \varphi \in H_o^{1,2}(B; I\!\!R^3),$$

and the result follows from Lemma 2.3.

$$\square$$

There is a local compactness condition related to Lemma 3.5:

Lemma 4.5:   Let $X_m = X(x_m) = X_{mo} + Z_m \in \mathcal{C}_o(\Gamma) + H_o^{1,2}(B; I\!\!R^3)$, with $x_m = (x_{mo}, z_m)$ satisfy the conditions

$$D(X_m) \leq c < \infty,$$
$$D_H(X_m) = E_H(x_m) \to \beta \ (m \to \infty),$$
$$g_H(x_m) \to 0 \ (m \to \infty).$$

Then the sequence $\{X_{mo}\}$ is strongly relatively compact in $\mathcal{C}_o(\Gamma)$ and a subsequence $\{X_m\}$ converges weakly to an $H-$surface $X = X(x) \in \mathcal{C}(\Gamma)$.

If

$$D_H(X) = E_H(x) > \beta - \frac{4\pi}{3H^2},$$

even the sequence $\{X_m\}$ itself is strongly relatively compact.

Proof of Lemma 4.5:   By Remark II.2.4 and Lemma II.2.8 we may assume that $x_m = (x_{mo}, z_m) \xrightarrow{w} x = (x_o, z)$ weakly, while $x_{mo} \to x_o$ uniformly on ∂B. Similarly, $X_m \xrightarrow{w} X$, $X_{mo} \xrightarrow{w} X_o, Z_m \xrightarrow{w} Z$ weakly, and $X_{mo} \to X_o$ uniformly in $\overline{B}$. Expanding, cp. the proof of Lemma II.2.11:

$$X_{mo} - X_o = \frac{d}{d\phi}\gamma(x_{mo}) \cdot (x_{mo} - x_o) - \int_{x_o}^{x_{mo}} \int_{x'}^{x_{mo}} \frac{d^2}{d\phi^2}\gamma(x'')dx''dx'$$

$$= \frac{d}{d\phi}\gamma(x_{mo}) \cdot (x_{mo} - x_o) + I_m \text{ on } \partial B$$

where

$$I_m \to 0 \text{ in } H^{1/2,2} \cap C^o(\partial B).$$

Moreover we note that

$$\int\limits_{B} \nabla \hat{Z} \nabla \hat{X} \, dw = 0, \quad \forall \hat{X} \in \mathcal{C}_o(\Gamma), \hat{Z} \in H_o^{1,2}(B; \mathbb{R}^3).$$

Hence we obtain (cp. the proof of Lemma 2.2):

$$
\begin{aligned}
2D(X_{mo} - X_o) &= \int\limits_{B} \nabla \left(X_m - X \right) \nabla \left(X_{mo} - X_o \right) \, dw \\
&= \int\limits_{B} \nabla X_m \nabla \left(X_{mo} - X_o \right) + 2H X_{m_u} \wedge X_{m_v} \cdot \left(X_{mo} - X_o \right) \, dw + o(1) \\
&= \int\limits_{B} \nabla X_m \nabla \left(h \left(\frac{d}{d\phi} \gamma(x_{mo}) \cdot (x_{mo} - x_o) \right) \right) \\
&\quad + 2H \, X_{m_u} \wedge X_{m_v} \cdot \left(h \left(\frac{d}{d\phi} \gamma(x_{mo}) \cdot (x_{mo} - x_o) \right) \right) \, dw + o(1) \\
&= \langle dE_H(x_m), \, (x_{mo} - x_o) \rangle + o(1) \le g_H(x_m) \, |x_{mo} - x_o| + o(1) \to 0.
\end{aligned}
$$

I.e. $X_{mo} \to X_o$ strongly in $H^{1,2}(B; \mathbb{R}^3)$, and $x_{mo} \to x_o$ strongly in $M^\dagger$.

By weak continuity of dD_H clearly

$$dD_H(X) = 0 \in \left(H_o^{1,2}(B; \mathbb{R}^3) \right)^*.$$

Hence

$$g_H(x) \le \sup_{\substack{y_o \in M^\dagger \\ |y_o - x_o| < 1}} \langle dE_H(x), x_o - y_o \rangle.$$

But for any $y_o \in M^\dagger$ Lemma 2.2 and strong convergence $x_{mo} \to x_o$:

$$
\begin{aligned}
\langle dE_H(x), x_o - y_o \rangle &= \int\limits_{\partial B} \partial_n X \frac{d}{d\phi} \gamma(x_o)(x_o - y_o) \, do \\
&= \lim_{m \to \infty} \int\limits_{\partial B} \partial_n X_m \frac{d}{d\phi} \gamma(x_{mo})(x_{mo} - y_o) \, do \\
&\le \lim_{m \to \infty} g_H(x_m) |x_{mo} - y_o| \to 0.
\end{aligned}
$$

This shows that x is critical, or equivalently, that X is an $H-$surface.

Finally, if $D_H(X) > \beta - \frac{4\pi}{3H^2}$, we may let $Y_m = X_o + Z_m = X_m - (X_{mo} - X_o) \in X_o + H_o^{1,2}(B; \mathbb{R}^3)$. Y_m satisfies $D_H(Y_m) \to \beta$, $dD_H(Y_m) \to 0 \in \left(H_o^{1,2}(B; \mathbb{R}^3) \right)^*$. Now we argue as in the proof of Lemma 3.5 to conclude that $Y_m \to X$ strongly. I.e. $X_m \to X$, and the proof is complete.

$$\square$$

In order to give the proof of Theorem 4.1. for simplicity we argue indirectly:

(4.2)   *Assume that there is no unstable $H-surface$ $X \in \mathcal{C}(\Gamma)$.*

Let
$$\beta_o = \inf\{D_H(X) \mid X \text{ is a relative minimum of } D_H \text{ on } \mathcal{C}(\Gamma)\}$$

By Theorem 4.2 $\beta_o \geq -c(\Gamma) > -\infty$.

Assumption (4.2) turns Lemma 4.5 into a local compactness condition comparable to Lemma 3.5:

(4.3)   For any $\beta < \beta_o + \frac{4\pi}{3H^2}$ any bounded sequence $X_m = X(x_m)$ with $x_m \in \mathcal{M}$ and $D_H(X_m) \to \beta$, $g_H(x_m) \to 0$ as $m \to \infty$ is relatively compact.

Define, as usual, for $\beta \in \mathbb{R}$
$$\mathcal{K}_\beta = \{x \in \mathcal{M} \mid E_H(x) = \beta, g_H(x) = 0\},$$
$$\mathcal{M}_\beta = \{x \in \mathcal{M} \mid E_H(x) < \beta\}.$$

Note that (4.2), (4.3) imply the following

Lemma 4.6: β_o is achieved at some relative minimum X_H of D_H on $\mathcal{C}(\Gamma)$. Moreover, for any β there exists a neighborhood $\mathcal{N}$ of $\mathcal{K}_\beta$ such that $\mathcal{N} \cap \mathcal{M}_\beta = \emptyset$.

Proof: A minimizing sequence $\{X_m = X(x_m)\}$ of relative minima of D_H in $\mathcal{C}(\Gamma)$ is bounded - by Theorem 4.2 - and hence relatively compact - by (4.3). Any accumulation point X_H by (4.2) must be a relative minimum.

By (4.2) again, for any β the set $\mathcal{K}_\beta$ is both relatively open and trivially closed in $\overline{\mathcal{M}_\beta}$. Hence, there exists a neighborhood $\mathcal{N}$ of $\mathcal{K}_\beta$ such that $\mathcal{N} \cap (\overline{\mathcal{M}_\beta} \backslash \mathcal{K}_\beta) = \emptyset$; a fortiori, $\mathcal{N} \cap \mathcal{M}_\beta = \emptyset$.

$\square$

Remark 4.7:   In the following we shall apply Lemma 4.6 for $\beta < \beta_o + \frac{4\pi}{3H^2}$ only. Note that by Theorem 4.2 for such β we may choose $\mathcal{N} \subset \{x \in \mathcal{M} \mid |x|_T < R_o\}$ where
$$R_o = \sup\{|x|_T \mid D(X(x)) \leq 3\beta_o + \frac{4\pi}{H^2} + c(\Gamma)\}.$$

An easy modification of the proof of Lemma II.1.9 also shows the following:

Lemma 4.8:   Suppose (4.3) is satisfied. Then for any $\beta < \beta_o + \frac{4\pi}{3H^2}$, any $\bar{\epsilon} > 0$, any neighborhood $\mathcal{N}$ of $\mathcal{K}_\beta$, any $R > 0$ there exists $\epsilon \in]0, \bar{\epsilon}[$, and a continuous deformation $\Phi : \mathcal{M} \times [0, 1] \to \mathcal{M}$ such that

i) $\Phi(x, t) = x$, if $t = 0$, or $|E_H(x) - \beta| \geq \bar{\epsilon}$, or $g_H(x) = 0$.

ii) $E_H(\Phi(x,t))$ is non-increasing in t, for any $x \in \mathcal{M}$.

iii) $\Phi(\mathcal{M}_{\beta+\epsilon} \cup \{x \in \mathcal{M} \mid |x|_T \geq R+1\}, 1) \subset \mathcal{M}_{\beta-\epsilon} \cup \mathcal{N} \cup \{x \in \mathcal{M} \mid |x|_T \geq R\}$.

Proof: Φ is obtained by integrating a pseudo-gradient vector field e cut off near the critical set. On a bounded region $\{x \in \mathcal{M} \mid |x|_T \leq R+1\}$ by (4.3) and boundedness of X all estimates from the proof of Lemma II.1.9 convey. Hence (iii) follows from $|e| = \left|\frac{\partial}{\partial t}\Phi\right| \leq 1$. (i) and (ii) are standard.

$\square$

By Lemma 3.4 given $X_H = X(x_H)$ with $x_H \in \mathcal{K}_{\beta_o}$ we can find $X_1 = X(x_1) \in \mathcal{C}(\Gamma)$ with $x_1 \in \mathcal{M}$ and
$$D_H(X_1) < D_H(X_H) .$$
Moreover,

(4.4) $P = \{p \in C^o\,([0,1];\, \mathcal{M}) \mid p(0) = x_H,\ p(1) = x_1\} \neq \emptyset$

and
$$\beta_H = \inf_{p \in P} \sup_{x \in p} E_H(x) < \beta_o + \frac{4\pi}{3H^2} \;.$$

More generally, for any relative minimum $\tilde{X} = X(\tilde{x}) \in \mathcal{C}(\Gamma)$, $\tilde{x} \in \mathcal{M}$, by convexity of $\mathcal{M}$
$$\tilde{P} = \{p \in C^o\,([0,1];\, \mathcal{M}) \mid p(0) = \tilde{x},\ p(1) = x_1\} \neq \emptyset$$
and we may let
$$\tilde{\beta} = \inf_{p \in \tilde{P}} \sup_{x \in p} E_H(x) .$$

For such an $\tilde{X}$ and $R > 0$ also introduce numbers
$$\tilde{\beta}^R = \inf_{p \in \tilde{P}} \sup_{\substack{x \in p \\ |x|_T \leq R}} E_H(x) \leq \tilde{\beta} .$$

Lemma 4.9   Suppose $D_H(\tilde{X}) < \beta_o + \frac{4\pi}{3H^2}$. Then for any $R \geq R_o + 1$ there holds the estimate
$$\tilde{\beta} \geq \tilde{\beta}^R > D_H(\tilde{X}).$$
R_o was defined in Remark 4.7.

Proof: Suppose by contradiction that for $R = R_o + 1$
$$\beta := \tilde{\beta}^R = D_H(\tilde{X}) < \beta_o + \frac{4\pi}{3H^2} \;.$$

Let $\bar{\epsilon} = D_H(\tilde{X}) - D_H(X_1) > 0$, and let $\mathcal{N}$ be a neighborhood of $\mathcal{K}_\beta$ as in Lemma 4.6 and Remark 4.7. Choose $\epsilon > 0$ and a deformation Φ according to Lemma 4.8, and let $p \in \tilde{P}$ satisfy
$$\sup_{\substack{x \in p \\ |x|_T \leq R}} E_H(x) < \beta + \epsilon.$$

By property i) of Φ the deformed path $p' = \Phi(p,1) \in \tilde{P}$. By iii), moreover,

$$p' \subset \mathcal{M}_{\beta-\epsilon} \cup \mathcal{N} \cup \{x \in \mathcal{M} \mid |x|_T \geq R_o\}.$$

Since $\mathcal{N} \cap \mathcal{M}_{\beta-\epsilon} = \emptyset$ by Lemma 4.6, while by Remark 4.7 $\mathcal{N} \cap \{x \in M \mid |x|_T \geq R_o\} = \emptyset$, we conclude that either $p' \subset \mathcal{N}$ or $p' \cap \mathcal{N} = \emptyset$. But $\tilde{x} \in p' \cap \mathcal{N}$ while $x_1 \in p' \backslash \mathcal{N}$.

I.e. p' intersects $\mathcal{N}$ but is not contained in $\mathcal{N}$. The contradiction proves the lemma.

$\square$

We now return to the case $\tilde{X} = X_H$. Note that by Theorem III.2.1 for $x = (x_o, z) \in \mathcal{M}$ uniformly bounded in $\mathcal{M}$ also $V(X(x))$ remains uniformly bounded. In consequence, the functional E_H is uniformly continuous in $H \in \mathbb{R}$ on any set $\{x \in \mathcal{M} \mid |x|_T \leq R\}$, and for $\overline{H}$ sufficiently close to our initially chosen H we have by Lemma 4.9:

$$\begin{aligned}
\beta_{\overline{H}} &:= \inf_{p \in P} \sup_{x \in p} E_{\overline{H}}(x) \\
&\geq \inf_{p \in P} \sup_{\substack{x \in p \\ |x|_T \leq R_o+1}} E_{\overline{H}}(x) > E_{\overline{H}}(x_H),
\end{aligned} \tag{4.5}$$

where P is defined by (4.4).

<u>**Lemma 4.10:**</u>   The map $\overline{H} \to \dfrac{\beta_{\overline{H}}}{\overline{H}}$ is non-increasing.

Proof: Use the identity for $0 < H_1 < H_2$ and $X \in \mathcal{C}(\Gamma)$:

$$\frac{D_{H_1}(X)}{H_1} - \frac{D_{H_2}(X)}{H_2} = \left(\frac{1}{H_1} - \frac{1}{H_2} \right) D(X) \geq 0. \tag{4.6}$$

Now suppose $H_1 < H_2$ are sufficiently close to H such that (4.5) holds. Let $p_m \in P$ be a minimizing sequence for H_1:

$$\sup_{x \in p_m} E_{H_1}(x) \to \beta_{H_1} \ (m \to \infty),$$

and let $x_m \in p_m$ satisfy

$$E_{H_2}(x_m) = \sup_{x \in p_m} E_{H_2}(x) \geq \beta_{H_2}.$$

Applying (4.6) with $X_m = X(x_m)$ we obtain that

$$\frac{\beta_{H_1}}{H_1} \geq \liminf_{m \to \infty} \frac{D_{H_1}(X_m)}{H_1} \geq \liminf_{m \to \infty} \frac{D_{H_2}(X_m)}{H_2} \geq \frac{\beta_{H_2}}{H_2}.$$

The lemma follows.

□

By a classical result in Lebesgue measure theorey, Lemma 4.10 implies that the map $\overline{H} \mapsto \frac{\beta_{\overline{H}}}{\overline{H}}$ is a.e. differentiable near H. Define

$$(4.7) \quad \mathcal{H} = \{\overline{H} \in I\!\!R \,|\beta_{\hat{H}} \text{ is defined near } \overline{H} \text{ and } \limsup_{\hat{H} \to \overline{H}} \left(\frac{\frac{\beta_{\hat{H}}}{\hat{H}} - \frac{\beta_{\overline{H}}}{\overline{H}}}{\overline{H} - \hat{H}} \right) < \infty \}.$$

$\mathcal{H}$ is dense in a neighborhood of H. Therefore, we may approximate H by numbers $H_m \in \mathcal{H}$, $H_m \to H$ $(m \to \infty)$. (If $H \in \mathcal{H}$, we may let $H_m \equiv H$.)

Still maintaining our assumption (4.2) for our initially chosen H we now establish:

Lemma 4.11: For any sufficiently large (fixed) $m \in I\!\!N$ the functional E_{H_m} satisfies the local Palais-Smale condition on $\mathcal{M}$:

Any sequence $\{X_m^k\}_{k \in N}$, $X_m^k = X(x_m^k), x_m^k \in \mathcal{M}$, with $D(X_m^k) \leq c$ uniformly, $D_{H_m}(X_m^k) = E_{H_m}(x_m^k) \to \beta_{H_m}$, $g_{H_m}(x_m^k) \to 0$ as $(k \to \infty)$ is relatively compact.

Proof: By Lemma 4.5 the thesis is true unless for some sequence $m \to \infty$ E_{H_m} admits a critical point x_m with $X_m = X(x_m)$ satisfying

$$D_{H_m}(X_m) \leq \beta_{H_m} - \frac{4\pi}{3H_m^2}.$$

By the construction of Lemma 3.4 we can estimate with a uniform constant β :

$$(4.8) \qquad \beta_{H_m} \leq \beta < \beta_o + \frac{4\pi}{3H^2}$$

for $m \geq m_o$. Moreover, the a-priori bound Theorem 4.2 guarantees that

$$D(X_m) \leq 3\beta + c(\Gamma) < \infty$$

for $m \geq m_o$. But then also

$$g_H(x_m) \to 0 \ (m \to \infty),$$

and by Lemma 4.5 $\{X_m\}$ weakly accumulates at an $H-$surface $X \in \mathcal{C}(\Gamma)$ with

$$D_H(X) \leq \liminf_{m \to \infty} D_{H_m}(X_m) < \beta_o, \text{ contradicting (4.2).}$$

□

Lemma 4.12:   For any sufficiently large $m \in I\!N$ there is a solution $X_m = X(x_m)$ of the Plateau problem (III.1.1)-(III.1.3) for H_m, characterized by the condition

$$D_{H_m}(X_m) = \beta_{H_m},$$

and x_m is a point of accumulation of a minimizing sequence of paths $p_m^k \in P, k \in I\!N$, such that

$$\sup_{x \in p_m^k} E_{H_m}(x) \to \beta_{H_m}(k \to \infty).$$

Proof: Fix $m \in I\!N$. Choose a sequence $\{H_m^k\}_{k \in I\!N}$ of numbers $H_m^k > H_m$, $H_m^k \to H_m \ (k \to \infty)$. Let $\{p_m^k\}_{k \in I\!N}$, $p_m^k \in P$ be a minimizing sequence for H_m such that

$$(4.9) \qquad\qquad \sup_{x \in p_m^k} E_{H_m}(x) \leq \beta_{H_m} + (H_m^k - H_m).$$

For arbitrary $x \in p_m^k$ with

$$(4.10) \qquad\qquad E_{H_m^k}(x) \geq \beta_{H_m^k} - (H_m^k - H_m)$$

by (4.6)-applied to $X = X(x)$-and (4.7) we obtain the uniform bound:

$$
\begin{aligned}
(4.11) \qquad D(X) &\leq \frac{H_m^k H_m}{H_m^k - H_m}\left(\frac{\beta_{H_m} + H_m^k - H_m}{H_m} - \frac{\beta_{H_m^k} - (H_m^k - H_m)}{H_m^k} \right) \\[2mm]
&\leq H_m^k H_m \left(\frac{\dfrac{\beta_{H_m}}{H_m} - \dfrac{\beta_{H_m^k}}{H_m^k}}{H_m^k - H_m} \right) + H_m^k + H_m \leq c < \infty.
\end{aligned}
$$

Suppose there exists $\delta > 0$ such that for all $x \in p_m^k$ satisfying (4.10) there holds

$$(4.12) \qquad\qquad g_{H_m^k}(x) \geq \delta > 0$$

uniformly in $k \in I\!N$.

By (4.11) and uniform continuity of E_H, g_H in H on bounded sets, for sufficiently large k a pseudo-gradient vector field for $E_{H_m^k}$ near such x will also be a pseudo-gradient vector field for E_{H_m} near x, and a pseudo gradient line deformation of p_m^k near points satisfying (4.10) will yield a sequence of comparison paths still satisfying (4.9).

So eventually (4.12) lets us arrive at a path $p' \in P$ where

$$\sup_{x \in p'} E_{H_m^k}(x) \leq \beta_{H_m^k} - (H_m^k - H_m) < \beta_{H_m^k},$$

contradicting the definition of $\beta_{H_m^k}$.

Negating (4.12), by (4.9) - (4.11) we find a sequence $\{X_m^k = X(x_m^k)\}$ such that

$$D(X_m^k) \leq c \,,$$

$$\beta_{H_m} \geq \liminf_{k \to \infty} E_{H_m}(x_m^k) = \liminf_{k \to \infty} E_{H_m^k}(x_m^k) \geq \liminf_{k \to \infty} \beta_{H_m^k} = \beta_{H_m} \,,$$

$$\lim_{k \to \infty} g_{H_m}(x_m^k) = \lim_{k \to \infty} g_{H_m^k}(x_m^k) \to 0 \ (k \to \infty),$$

$$x_m^k \in p_m^k.$$

By Lemma 4.11 $\{x_m^k\}$ accumulates at a critical point x_m of E_{H_m}.

$\square$

Proof of Theorem 4.1:   For $H_m \in \mathcal{H}$ tending to H let $X_m = X(x_m)$ be the solutions obtained in Lemma 4.12. By Theorem 4.2

$$D(X_m) \leq c < \infty$$

while by (4.8) we may assume that

$$\beta_{H_m} = E_{H_m}(x_m) \to \beta < \beta_o + \frac{4\pi}{3H^2},$$

and

$$g_H(x_m) \to 0.$$

By Lemma 4.5, assumption (4.2), and the definiton of β_o, the sequence $\{x_m\}$ is relatively compact and accumulates at a critical point $\tilde{x} \in \mathcal{M}$ of E_H. Moreover, $\tilde{x}$ is an accumulation point of paths $p_m \in P$ where

(4.13) $$\sup_{x \in p_m} E_{H_m}(x) \to E_H(\tilde{x})$$

If $\tilde{X} = X(\tilde{x}) \in \mathcal{C}(\Gamma)$ were a relative minimum of D_H, now (4.13) would give a contradiction to Lemma 4.9. Hence (4.2) cannot be true, and the proof is complete.

$\square$

Finally we present the proof of Theorem III.3.4. Recall the assertion:

Theorem 4.13:   If Γ is a Jordan curve of class C^2 in $\mathbb{R}^3$, $H \in \mathbb{R}$, and if for some $\hat{X} \in \mathcal{C}(\Gamma)$ there holds

$$H^2 D(\hat{X}) < \frac{2}{3}\pi,$$

then D_H admits a relative minimum X_H on $\mathcal{C}(\Gamma)$ characterized by the conditions that

$$D(X_H) < 5D(\hat{X}),$$
$$D_H(X_H) = \min \left\{ D_H(X) \mid X \in \mathcal{C}(\Gamma),\ D(X) < 5D(\hat{X}) \right\}.$$

Proof of Theorem 4.13:   By Theorem I.4.10 we may asssume that $\hat{X}$ is a minimal surface. Moreover, it remains to consider the case $H \neq 0$.

Let

$$\hat{\mathcal{M}} = \left\{ x \in \mathcal{M} \mid D(X(x)) < 5D(\hat{X}) \right\}.$$

Define

$$\beta_o = \inf_{x \in \hat{\mathcal{M}}} E_H(x).$$

Claim 1: $\beta_o > -\infty.$

Let $x \in \hat{\mathcal{M}}$, $X = X(x)$. Applying a variant of the isoperimetric inequality, (cp. Remark III.2.2.ii), we may estimate

$$|V(X)| \leq |V(\hat{X})| + \left[\frac{1}{36\pi} \left(D(X) + D(\hat{X}) \right)^3 \right]^{1/2} \leq c < \infty,$$

uniformly, and the claim follows.

Claim 2: $\beta_o < \inf \left\{ D_H(X) \in \mathcal{C}(\Gamma),\ D(X) = 5D(\hat{X}) \right\} =: \hat{\beta}$

Simply estimate, using the isoperimetric inequality

$$D_H(X) - D_H(\hat{X}) = D(X) - D(\hat{X}) + 2H(V(X) - V(\hat{X}))$$

$$\geq 4D(\hat{X}) - 2|H|\sqrt{\frac{6D(\hat{X})^3}{\pi}}$$

(4.14)

$$= 4D(\hat{X}) \left(1 - \sqrt{\frac{3H^2 D(\hat{X})}{2\pi}} \right) \geq 0,$$

if $D(X) = 5D(\hat{X})$.

Since $\hat{X}$ is a minimal surface, while $H \neq 0$, it follows that

$$dD_H(\hat{X}) \neq 0 \in H_o^{1,2}(B; I\!R^3)^*$$

and there exists a surface $\tilde{X} \in \hat{X} + H_o^{1,2}(B; I\!R^3)$ such that $D(\tilde{X}) < 5D(\hat{X})$ and

$$\beta_o \leq D_H(\tilde{X}) < D_H(\hat{X}) \leq \hat{\beta}.$$

Now remark that Lemma 4.5 implies that any sequence $\{x_m\} \subset \hat{\mathcal{M}}$ such that

$$E_H(x_m) \to \beta_o,$$
$$g_H(x_m) \to 0$$

is relatively compact, i.e. E_H satisfies the Palais-Smale condition $(P.S.)_{\beta_o}$ on $\hat{\mathcal{M}}$.

Indeed, by Lemma 4.5 we may assume that $X_m = X(x_m) \xrightarrow{w} X_o = X(x)$. By weak lower semi - continuity of Dirichlet's integral X satisfies $D(X) \leq 5D(\hat{X})$. In particular, $x \in \hat{\mathcal{M}}$, $E_H(x) \geq \beta_o$, and by Lemma 4.5 $X_m \to X$ strongly as $m \to \infty$.

Finally, suppose by contradiction that β_o is a regular value of E_H on $\hat{\mathcal{M}}$. By $(P.S.)_{\beta_o}$ there exists $\delta_o > 0$ such that

(4.15) $$g_H(x) \geq \delta_o$$

for $x \in \hat{\mathcal{M}}$ with $E_H(x) \leq \beta_o + \delta_o$.

For $\delta > 0$ let

$$\hat{\mathcal{M}}_\delta = \left\{ x \in \hat{\mathcal{M}} \mid E_H(x) < \beta_o + \delta \right\},$$

and let $\hat{e} : \hat{\mathcal{M}}_\delta \to T$ be a Lipschitz continuous pseudo-gradient vector field for E_H on $\hat{\mathcal{M}}_{\delta_o}$ satisfying the conditions

$$\hat{e}(x) + x \in \mathcal{M},$$
(4.16) $$|\hat{e}(x)|_T < 1$$
$$\langle dE_H(x), \hat{e}(x) \rangle < -\min\{\frac{1}{2} g_H^2(x), 1\},$$

which we may construct according to Lemma II.1.8.

Let

$$\hat{\Phi} : \mathcal{D}(\hat{\Phi}) \subset \hat{\mathcal{M}}_{\delta_o} \times [0, 1] \to \hat{\mathcal{M}}_{\delta_o}$$

be a maximal flow of integral curves of $\hat{e}$, cp. Lemma II.1.9. Note that for any $\delta < \min\{\delta_o, \hat{\beta} - \beta_o\}$ the set $\hat{\mathcal{M}}_\delta$ is forwardly invariant under $\hat{\Phi}$, i.e.

(4.17) $$\hat{\Phi}\left(\hat{\mathcal{M}}_\delta, t\right) \subset \hat{\mathcal{M}}_\delta, \ \forall t \geq 0.$$

Hence for such δ clearly $\mathcal{D}(\hat{\Phi}) \supset \hat{\mathcal{M}}_\delta \times [0,1]$.

But by (4.15), (4.16) for any $x \in \hat{\mathcal{M}}_\delta$

$$E_H\left(\hat{\Phi}(x,1)\right) = E_H(x) + \int_0^1 \langle dE_H(\hat{\Phi}(x,t)),\ \hat{e}(\cdot)\rangle\, dt$$

$$< \beta_o + \delta - \min\{\frac{1}{2}\,\delta_o^2, 1\} < \beta_0$$

if $\delta < \frac{\delta_o^2}{2}$. In view of (4.17) this contradicts the definition of β_o, and the theorem follows.

□

Remark 4.14: Inspection of the above proof shows that (4.14) and Claim 2 remain true if $H^2 D(\hat{X}) \leq \frac{2}{3}\pi$. By uniform boundedness of the volume $V(X(x))$, $x \in \hat{\mathcal{M}}$, cp. Claim 1, $D_H(X(x))$ is uniformly continuous as a function of H on $\hat{\mathcal{M}}$. Hence the estimate $\beta_o < \hat{\beta}$ of Claim 2, and thus also Theorem 4.13 will stay true for a sligthly larger range of curvatures H.

References:

Adams, R.A.:   [1]   Sobolev spaces,
Academic Press, New York - San Francisco -
London, 1975

Almgren, F.:   [1]   Plateau's Problem. An invitation to varifold
geometry, Benjamin, New York - Amsterdam,
1966

Almgren, F. - Simon, L.:   [1]   Existence of embedded solutions of Plateau's
Problem, Ann. Sc. Norm. Sup. Pisa Cl. Sci.
(4) 6 (1979), 447-495

Ambrosetti, A. - Rabinowitz, P.H.:
                         [1]   Dual variational methods in critical point
theory and applications, J. Functional
Analysis 14 (1973), 349 - 381

Alt, W.:   [1]   Verzweigungspunkte von H-Flächen,
Part I: Math. Z. 127 (1972), 333-362
Part II: Math. Ann. 201 (1973), 33-55

Birkhoff, G.D.:   [1]   Dynamical systems with two degrees of free-
dom, Trans. AMS, 18 (1917), 199-300

Böhme, R.:   [1]   Die Jacobifelder zu Minimalflächen im $I\!R^3$,
Manusc. Math. 16 (1975), 51-73

Böhme, R. - Tromba, A.J.:   [1]   The index theorem for classical minimal surfaces, Ann. of Math. 113 (1981), 447 - 499

Brezis, H. - Coron, J.-M.:   [1]   Sur la conjecture de Rellich pour les surfaces à
courbure moyenne prescrite.
C.R. Acad. Sci. Paris Ser. I. 295 (1982),
615 - 618
                         [2]   Multiple solutions of H-systems and Rellich's
conjecture, Comm. Pure Appl. Math. 37 (1984),
149 - 187

Chen, Y.W.:   [1]   Branch points, poles, and planar points of minimal surfaces in $I\!R^3$, Ann. Math. (2) 49 (1948), 790-806

Conley, C.:   [1]   Isolated invariant sets and the Morse index, CBMS 38, AMS, Providence, 1978

Conley, C. - Zehnder, E.:   [1]   A Morse type index theory for flows and periodic solutions for Hamiltonian systems, Comm. Pure Appl. Math. 37 (1984), 207 - 253

Courant, R.:   [1]   Dirichlet's principle, conformal mapping and minimal surfaces, New York, Interscience, 1950
                                   [2]   Plateau's Problem and Dirichlet's principle, Ann. of Math. 38 (1937), 679 - 724
                                   [3]   Critical points and unstable minimal surfaces Proc. Nat. Acad. Sci. USA 27 (1941), 51-57

Douglas, J.:   [1]   Solution of the Problem of Plateau, Trans. AMS 33 (1931), 263 - 321

Djugundji, J.:   [1]   Topology, Allyn and Bacon, Boston, 1966

Gilbarg, D. - Trudinger, N.S.:
                                   [1]   Elliptic partial differential equations of second order, Springer, Grundlehren 224, Berlin, 1977

Gulliver, R.D.:   [1]   Regularity of minimizing surfaces of prescribed mean curvature, Ann. Math. (2) 97 (1973), 275-305
                                   [2]   to appear in Concus, P. - Finn, R. (eds.): Variational methods in free surface interfaces, Springer

Gulliver, R.D. - Lesley, F.D.: [1]   On boundary branch points of minimizing surfaces, Arch. Rat. Mech. Anal. 52 (1973), 20-25

Gulliver, R.D. - Ossermann, R. - Royden, H.L.:
                                   [1]   A theory of branched immersions of surfaces, Amer. J. Math. 95 (1973), 750-812

Hartmann. P. - Wintner, A.: [1]   On the local behavior of solutions of non-para-
bolic partial differential equations,
Amer. J. Math. 75 (1953), 449-476

Heinz, E.:   [1]   Über die Existenz einer Fläche konstanter
mittlerer Krümmung bei vorgegebener
Berandung, Math. Ann. 127 (1954), 258 - 287
[2]   On the nonexistence of a surface of constant
curvature with finite area and prescribed
rectifiable boundary,
Arch. Rat. Mech. Anal. 35 (1969), 249 - 252
[3]   An inequality of isoperimetric type for surfaces
of constant mean curvature,
Arch. Rat. Mech. Anal. 33 (1969), 155-168
[4]   Ein Regularitätssatz für Flächen beschränkter
mittlerer Krümmung,
Nachr. Akad. Wiss. Göttingen (1969), 107-118
[5]   Unstable surfaces of constant mean curvature,
Arch. Rat. Mech. Anal. 38 (1970), 256-267

Hildebrandt, S.:   [1]   Boundary behavior of minimal surfaces,
Arch. Rat. Mech. Anal. 35 (1969), 47 - 82
[2]   On the Plateau problem for surfaces of constant
mean curvature,
Comm. Pure Appl. Math.23 (1970), 97 - 114
[3]   Über Flächen konstanter mittlerer Krümmung,
Math. Z. 112 (1969), 107-144
[4]   The calculus of variations today, as reflected in
the Oberwolfach meetings,
Perspect. Math., Birkhäuser, 1985

Jäger, W.:   [1]   Ein Maximumsprinzip für ein System nichtli-
nearer Differentialgleichungen,
Nachr. Akad. Wiss. Göttingen 11 (1976), 175-164

Kelly, J.L.:   [1]   General topology,
Springer, Berlin - Heidelberg - New York, 1975

Ladyshenskaya, O.A. - Ural'ceva, N.N.:
[1]   Linear and quasilinear elliptic equations,
Academic Press, New York, 1968

Lebesgue, H.:   [1]   Sur le problème de Dirichlet,
Rend. Circ. Mat. Palermo 24 (1907), 371-402

Ljusternik, L.A. - Schnirelmann, L.:

[1] Existence de trois géodésics fermés sur des surfaces de genre 0,
C.R. Acad. Sci. Paris 188 (1929), 534-536

Meeks, W.H. - Yau, S. T.: [1] The classical Plateau Problem and the topology of three-dimensional manifolds,
Topology 21 (1982), 409-442

Milnor, J.: [1] Morse theory,
Ann. Math. Studies 51 (1963)

Morrey, C.B.: [1] Multiple integrals in the calculus of variations,
Grundlehren 130, Berlin- Heidelberg - New York, 1966

[2] The problem of Plateau on a Riemannian manifold, Ann. Math. 49 (1948), 807-851

Morse, M. - Tompkins, C.B.: [1] The existence of minimal surfaces of general critical types,
Ann. of Math. 40 (1939), 443 - 472

Nitsche, J.C.C.: [1] Vorlesungen über Minimalflächen
Grundlehren 199, Springer Verlag, Berlin - Heidelberg - New York, 1975

[2] The boundary behavior of minimal surfaces. Kellog's theorem and branch points on the boundary,
Invent. Math. 8 (1969), 313-333;
Addendum, Invent. Math. 9 (1970), 270

[3] A new uniqueness theorem for minimal surfaces
Arch. Rat. Mech. Anal 52 (1973), 319 - 329

Osserman, R.: [1] A proof of the regularity everywhere of the classical solution to Plateaus's problem,
Ann. Math. 91 (1970), 550-569

Palais, R.S.:   [1]   Morse theory on Hilbert manifolds,
Topology 2 (1963), 299 - 340

[2]   Lusternik - Schnirelman theory in Banach manifolds,
Topology 5 (1966), 115 - 132

[3]   Critical point theory and the minimax principle,
Proc. Symp. Pure Appl. Math. 15 (1970), 185-212

Palais, R.S. - Smale, S.:   [1]   A generalized Morse theory,
Bull. AMS 70 (1964), 165 - 172

Pucci, P. - Serrin, J.:   [1]   Extensions of the mountain-pass theorem,
J. Funct. Anal. 59 (1984), 185 - 210

Radó, T.:   [1]   On Plateau's problem,
Ann. of Math. 31 (1930), 457 - 469

[2]   The problem of least area and the problem of Plateau,
Math. Z. 32 (1930), 763-796

[3]   Contributions to the theory of minimal surfaces,
Acta Litt. Sci. Univ. Szeged 6 (1932), 1-20

[4]   The isoperimetric inequality and the Lebesgue definition of surface area,
Trans.AMS 61 (1947), 530-555

Rybakowski, K.P.:   [1]   On the homotopy index for infinite - dimensional semiflows,
Trans. AMS 269 (1982), 351-382

Rybakowski, K.P. - Zehnder, E.:
[1]   On a Morse equation in Conley's index theory for semiflows on metric spaces,
Ergodic Theory Dyn. Syst. 5 (1985), 123-143

Sasaki, R.:   [1]   On the total curvature of a closed curve,
Jap. J. Math. 29 (1959), 118-125

Schwarz, H.A.:   [1]   Gesammelte Mathematische Abhandlungen I,
Springer, Berlin, 1890

Schüffler, K. - Tomi, F.:   [1]   Ein Indexsatz für Flächen konstanter mittlerer Krümmung,
Math. Z. 182 (1983), 245-257

Serrin, J.:   [1]   On surfaces of constant mean curvature which
                                      span a given space curve,
                                      Math. Z. 112 (1969), 77-88

Simon, L.:   [1]   Lectures on geometric measure theory,
                                      Canberra (1985)

Shiffman M.:   [1]   The Plateau problem for non - relative minima,
                                      Ann. of Math. 40 (1939), 834 - 854

Smale, S.:   [1]   Morse theory and a nonlinear generalization of
                                      the Dirichlet problem,
                                      Ann. of Math. (2) 80 (1964), 382 - 396

Spanier, H.E.:   [1]   Algebraic topology,
                                      McGraw Hill, New York, 1966

Söllner, M.:   [1]   Plateau's problem for surfaces of constant
                                      mean curvature from a global point of view,
                                      Manusc. Math. 43 (1983), 191-217

Steffen, K.:   [1]   Flächen konstanter mittlerer Krümmung mit
                                      vorgegebenem Volumen oder Flächeninhalt,
                                      Arch. Rat. Mech. Anal. 49 (1972), 99 - 128
                               [2]   Ein verbesserter Existenzsatz für Flächen kon-
                                      stanter mittlerer Krümmung,
                                      Manuscripta Math. 6 (1972), 105 - 139
                               [3]   On the existence of surfaces with prescribed
                                      mean curvature and boundary,
                                      Math. Z. 146 (1976), 113 - 135
                               [4]   On the nonuniqueness of surfaces with prescrib-
                                      ed constant mean curvature spanning a given
                                      contour,
                                      Arch. Rat. Mech. Anal. 94 (1986), 101-122

Steffen, K. - Wente, H.C.:   [1]   The non-existence of branch points in a solution
                                      to certain classes of Plateau type variational
                                      problems,
                                      Math. Z. 163 (1978), 211-238

Ströhmer, G.:   [1]   Instabile Flächen vorgeschriebener mittlerer
                                      Krümmung,
                                      Math. Z. 174 (1980), 119-133

Struwe, M.:   [1]   On a critical point theory for minimal surfaces
                                      spanning a wire in $\mathbb{R}^n$,
                                      J. Reine Angew. Math. 349 (1984), 1 - 23
                                [2]   Large H-surfaces via the mountain - pass - lemma,
                                      Math. Ann. 270 (1985), 441-459
                                [3]   Nonuniqueness in the Plateau problem for sur-
                                      faces of constant mean curvature,
                                      Arch. Rat. Mech. Anal. 93 (1986), 135-157

Thiel, U.:   [1]   The index theorem for k−fold connected
                                      minimal surfaces,
                                      Math. Ann. 270 (1985), 489-501
                                [2]   On the stratification of branched minimal sur-
                                      faces,
                                      preprint

Tomi, F.:   [1]   On the local uniqueness of the problem of least
                                      area,
                                      Arch. Rat. Mech. Anal. 52 (1973), 312-318

Tomi, F. - Tromba, A.J.:   [1]   Extreme curves bound embedded minimal sur-
                                      faces of the type of the disc,
                                      Math. Z. 158 (1978), 137-145

Tromba, A.J.:   [1]   On the number of simply connected minimal sur-
                                      faces spanning a curve,
                                      Memoirs AMS 194 (1977)
                                [2]   Degree theory on oriented infinite dimensional
                                      varieties and the Morse number of minimal
                                      surfaces spanning a curve in $\mathbb{R}^n$,
                                      Part I: $n \geq 4$, Trans. AMS 290 (1985), 385-413
                                      Part II: $n = 3$, Manusc. Math. 48 (1984), 139-161

Wente, H.C.:   [1]   An existence theorem for surfaces of constant
                                      mean curvature,
                                      J. Math. Anal. Appl. 26 (1969), 318 - 344
                                [2]   A general existence theorem for surfaces of con-
                                      stant mean curvature,
                                      Math. Z. 120 (1971), 277 - 288
                                [3]   The Dirichlet problem with a volume constraint,
                                      Manusc. Math. 11 (1974), 141-157
                                [4]   The differential equation $\Delta x = 2Hx_u \wedge x_v$
                                      with vanishing boundary values,
                                      Proc. AMS 50 (1975), 59 - 77
                                [5]   Large solutions to the volume constrained Pla-
                                      teau problem,
                                      Arch. Rat. Mech. Anal. 75 (1980), 59 - 77
                                [6]   Counterexample to a conjecture of H. Hopf,
                                      Pacific J. Math. 121 (1986), 193 -243

Werner, H.:   [1]   Das Problem von Douglas für Flächen konstanter
                                      mittlerer Krümmung,
                                      Math. Ann.133 (1957), 303 - 319

Widman, K.O.:   [1]   Hölder continuity of solutions of elliptic systems,
                                      Manusc. Math. 5 (1971), 299-308